Glossary of Applied Management and Financial Statistics

 W9-DIM-063

Glossary of Applied Management and Financial Statistics

E. J. Broster

Gower Press

First published in Great Britain by Gower Press Limited, Epping, Essex, 1974

© E.J. Broster 1974

ISBN 0 7161 0174 2

Robert Manning Strozier Library

DEC 4 1974

Tallahassee, Florida

Set in 11/13 Press Roman at Grosvenor Press
Printed in Great Britain
at the University Printing House, Cambridge
(Brooke Crutchley, University Printer)

Contents

Preface

To most business managers, the problem comes first, and the means of solving it second. Seeking a statistical technique or formula, or a source of environmental data can be a time-consuming exercise. The object of this glossary is to provide managers at all levels with a ready means of finding the technique or the data they may require in the solution of the numerical problems that come their way. The glossary includes a number of articles defining and demonstrating the statistical methods which are of the greatest use to management.

References are arranged in alphabetical order, and are numbered consecutively within their initial letters. This code is designed to facilitate cross-reference and, in particular, reference from an alphabetical index of applications at the end. There is also a short bibliography with comments.

Statistical theory has been avoided as far as possible with the result that some of the algebraic formulae given in the glossary have to be taken on trust. The theory underlying these formulae and mathematical proofs will be found in the appropriate works listed in the bibliography. Proofs of some of the algebraic formulae given, such as those of actuarial arithmetic, are simple enough and will all be found in any good elementary textbook on algebra.

Scope is always a problem with glossary compiling. The compiler may easily go too far or not far enough. As it is impossible to steer a middle course, one that satisfies everybody, it is hoped that, since going too far is the lesser of the two evils, this glossary errs on the liberal side rather than the other.

Once the question of scope has been settled, it is equally difficult to decide what items fall within that scope and what must be rejected as falling outside, what are relevant to the field and what are not.

It has not always been easy to decide whether a particular item belonged to the glossary or to the index of applications. Decision-tree synthesis and interpretation is a case in point. This explains why *Decision tree* and some other items appear in both glossary and index.

E.J.Broster

A1 Abscissa
The horizontal axis of a graph; often called the x-axis (Q1).

A2 Absolute measure
A value expressed in terms of units of quantity or of money, as distinct from relative measure (R7).

A3 Absorption costing
The costing system of accounting for overheads by allocating them to the products and services of the company; accomplished by making percentage additions to the direct labour content or direct materials content of each item. The percentages are subject to revision annually or more frequently. There are now many critics of the system who variously describe it as a folly, uneconomic, misleading and a waste of time and money. It forms the basis of the cost-plus method of pricing, but both the system and the method are tending to fall into disrepute in favour of marginal costing (M4) and optimum pricing (R6).

A4 Activity index
A percentage relative (R10) or index number (I7) of activity used in budgeting (B16). Some annual costs vary with the actual output, others, such as certain overheads, with the planned output. Some costs vary with the actual input of labour or materials, others with the planned input. A full-blooded system of activity indices has a separate activity index for each item of annual cost, apart, possibly, from items such as directors' fees.

Applications. Functional budgeting, planning, cost control.

A5   Addition law

A law of chance, which, when there are two or more exclusive events, calls for the addition of the chances of success to give the chance of one success. A punter decides to back the first and second-favourites in a race. The starting prices are 5 to 4 on and 2 to 1 against. Based on the betting market reckoning, the chance that one of the two will win is

$$\frac{5}{9} + \frac{3}{9} = \frac{8}{9} = 0.89$$

or 8 to 1 on

Applications. Merger and take-over arithmetic, investment appraisal.

A6   Advertising expenditure of companies

Estimates are given quarterly, in *The Statistical Review of Advertising,* of the expenditure on advertising by many producers and distributors of consumer goods; this covers all the principal media. There are no statistics of brand advertising. Nevertheless, company statistics may often be taken as giving a rough guide to brand advertising. They can be used in index form as a factor in the demand model (D7).

A7   Aggregative analysis

The analysis of aggregates, such as annual costs and sales, as distinct from marginal analysis (M3).

A8   Amount of £

The sum which a £ invested now at a given rate of compound interest will amount to in a given number of years' time, ie

$$A = (1 + \frac{R}{100})^n$$
$$= (1 + r)^n$$

where A is the amount, R the rate of interest per cent, r the rate of interest per £, and n the number of years. Printed tables of the amount of £ are published (A9).

Applications. Investment appraisal, pension-fund calculations.

A9   Amount of £ a year

The sum which a £ invested at compound interest at the end of each year will amount to in a given number of years' time. It is equal to the sum of the amounts of £ up to the end of the previous year plus £1, the formula being

$$A \text{ of } £ \text{ a year} = \frac{(1 + r)^n - 1}{r}$$

where r is the rate of interest per £, and n is the number of years. Printed tables of the amount of £ a year for a range of interest rates are published in *Inwood's Tables of Interest and Mortality* and *Parry's Valuation Tables*.

Applications.   Investment appraisal, pension-fund calculations.

A10   Analysis of variance

The analysis of a difference between two figures relating to the same phenomenon (such as between an estimate and an actual figure) by means of the causes of that difference. It has two connotations, one statistical and the other management accounting, which are considered below.

Statistical connotation.   The extent to which the variance equal to the square of the standard deviation (S24), ie $\Sigma (x^2)/(n-1)$, of a sample series of figures can be said to be attributable to the variance in the universe from which the sample is drawn, and the extent to which it can be said to be attributable to bias within the sample. The object of the exercise is to determine the representativeness (R13) of the sample. No sample can be expected to be fully representative of its universe (U3), but where the variance due to bias is relatively small, it can be accepted as reasonably representative. A fully representive sample has the same average and standard deviation as those of the universe. See F test (F10).

Applications.   Sampling.

Management-accounting connotation.   The extent to which the difference between, eg actual cost or actual sales revenue and the standard cost (S23) or the budgeted (B15) sales revenue can be attributed to each of a number of causes of the difference. For example, the actual cost of raw materials consumed in the period may be more, or less, than the standard cost. The shortfall may be due to a lower consumption or lower prices, or to both, or to a much lower consumption partly offset by higher prices; and so on.

Applications.   Budgeting and cost, sales and net-revenue analysis.

A11   Annual capital charge (ACC)

A capital outlay converted to an annual cost, so that it can be added to annual operating costs without violating the rules of arithmetic. There are many methods of conversion, ranging from the crude to the logically sophisticated. For the most acceptable method, see A12.

Applications.   Investment appraisal, annual costing.

A12   Annual value (AV)

One of the so-called investment criteria, probably the most sensible, logical and realistic. It converts the capital outlay to an annual capital charge (A11) by charging interest at the firm's borrowing rate on the total outlay, and depreciation, by sinking fund (S18) at the lending rate of interest on each individual wasting fixed asset. For other criteria, see I14, N4, T3, T15.

Applications.   Investment appraisal.

A13   Annuity

A limited annuity such as a life annuity is an annual payment made in equal instalments for a given number of years, or for life, on payment of a capital sum. The annuity consists of interest and capital repayments. In actuarial practice it is usually expressed as the annuity which a £ will purchase, for which the formula is the reciprocal of the present value of £ a year (P23), ie

$$\text{Limited annuity which £ will purchase} = \frac{r}{1-(1+r)^{-n}}$$

where r is the rate of interest per £, and n is the number of years.

A perpetual annuity, generally referred to as a *perpetuity,* is the simple-interest yield of a capital sum invested in an undated security. For deferred annuities and perpetuities, see D5.

Applications.   Investment appraisal, pension-fund calculations.

A14   Arithmetic mean (AM)

Also known as the arithmetic average, the simplest and most useful of all averages. The AM of a series of figures is their sum divided by their number:

$$AM = \frac{\Sigma X}{n}$$

where Σ is the sign of summation, X the figures in the series, and n the number of figures in the series. It is of wide and general application in all fields.

A15   Attribution costing

A product-costing system in which all costs directly attributable to the production and distribution of the product are ascribed to the product. It covers the marginal cost and fixed or time costs arising in respect of buildings, machinery, plant or vehicles used exclusively for the product. Where an establishment of a company is entirely given over to the production and distribution of a single brand, the whole of the annual costs incurred at the establishment, including the salaries of the establishment management team, are attributable to the brand. Where there are two or more brands produced at an establishment, it is not usually possible to go beyond the production and distribution capacity for attribution. If the units of production capacity specific to a particular brand are all the same in respect of annual costs, the total attributable annual cost of the brand may be expressed in the following equation:

$$T_a = aQ + bn$$

where T_a is the total attributable annual cost, a the marginal cost (M4) of the brand, Q the annual quantity of output, b the annual cost of providing and maintaining a unit of capacity, and n the number of units of capacity.

Attribution costing is designed to be realistic. It goes some way towards accounting for overhead costs, but to avoid arbitrary allocations, as under a system of absorption costing (A3), it necessarily does not go all the way.

Applications. Product variable costing, investment appraisal, efficiency auditing.

A16   Average

A generic term for all types of average; variously described as a measure of location or a measure of central tendency. It is sometimes called the *mean.* Where used unqualified, the term *average* can be taken to refer to the arithmetic mean (A14). Also see G2, H1, M14 and 15. For moving averages and weighted average, see M21 and W2.

A17   Average deviation

Sometimes called the *mean deviation*, it is a simple measure of dispersion (D19) of a series of figures from their arithmetic mean. It is equal to the sum of the deviations from the mean, with sign ignored, divided by the number of figures in the series.

$$\text{Average deviation} = \frac{\Sigma x}{n}$$

where Σ is the sign of summation, x the deviation from the mean, and n the number of figures in the series. Statisticians prefer the *standard* deviation (S24). Has wide and general application in all fields except statistical theory.

B1   Bar chart
A form of diagrammatic presentation used for comparing different magnitudes. It consists of two or more bars of equal width, usually upright, showing the make-up of some entity at different times or in different places. Each constituent of the bars has its own colour or its own form of shading, thus making a visual comparison easy. The lengths of the bars and the areas of the constituent parts are *pro rata* to the absolute magnitudes of the statistics that they represent.

B2   Basic statistics
The raw materials of statisticians; usually collated, but rarely processed beyond the stage of correction for extraneous factors, such as differences in materials prices and wage rates where the basic statistics consist of annual costs. They can be classified in various ways, one being the dichotomy of continuous series and discrete series (C21, D17); another dichotomy is that of time series and cross-section series (T8, C37), and a third, that of internal and environmental statistics.

B3   Bell-shaped curve
A symmetrical frequency-distribution curve (F9).

B4   Bessel's correction
A factor by which the variance (A10) of a small sample is multiplied to provide an estimate of the variance of the universe (U3). The correction factor is
$$\frac{n}{n-1}$$

where n is the number of items in the sample. The greater the value of n, the closer the correction approaches to unity, so that for very large samples the correction becomes unnecessary. See sampling (S3).

B5   Best fit, line of

A straight line or curve fitted as a graph representing the relationship between two variables plotted one against the other in a scatter diagram (S9). Figure B5.1 shows a scatter diagram of monthly costs plotted against monthly output. The line of best fit may be drawn freehand by personal judgement, as it is in Figure B5.1; or it may be fitted by group averages (G9) or least squares (L1), or by any other method of simple-regression analysis (R7). When the line has been drawn, then if it turns out to be straight, readings can be taken from it in order to derive its equation. If it turns out to be a curve, then it would be necessary for analytical purposes to determine the type of equation to which it conforms, eg logarithmic or quadratic. See model building (M16).

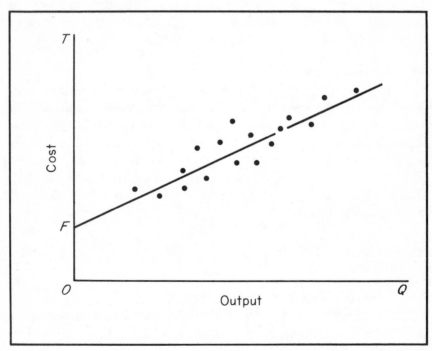

Figure B5.1   Scatter diagram showing line of best fit

B6   Beta coefficient (β)

A standardized measure of a regression coefficient (R8); achieved by 'stating each of the variables in units of their own individual standard deviation'. A simple regression equation (R9) may be written:

$$Y = aX + b$$

Its beta equivalent is:

$$\frac{Y}{\sigma_y} = \beta \frac{X}{\sigma_x} + b^1$$

Each figure of the Y series is divided by the standard deviation of the Y series, and each figure of the X series is divided by the standard deviation of the X series, before the regression analysis is carried out. The standard deviation of a series expressed in units of cwt is 20 times that of the same series expressed in units of a ton; the resultant ratios are the same whether the series is expressed in cwts or tons. At the same time the size of any regression coefficient depends partly on the units in which the basic series are expressed. The beta coefficient eliminates this factor from the regression, and the several beta coefficients of a multiple-regression equation can be compared directly one with another, and with the beta coefficients of other regression equations. For standard deviation, see S24.

B7   Betterment

Net fixed capital formation (F7).

B8   Bias

Lopsidedness. Non-random samples are said to have bias. It is often difficult, and sometimes impossible, to obtain a random sample (R3). Many samples are necessarily self-selected. Our parliamentary democracy itself is open to criticism in this respect. People who stand for election to Parliament or to local government are different from those who do not, ie although they may be representatives of the public, they are not necessarily representative of the public. They are largely self-selected. A parliament or local council truly representative of the public would consist of a random selection of the electorate. This not a matter of political colour. A random selection would probably give much the same kind of relative party strength as from an election. There may be serious differences in outlook however: in other words, there may be bias. There is evidence for supposing that small district councils vary appreciably in their bias.

Some are more uncharitable than others in the way that they deal with ratepayers' problems; some are more high-handed than others; some always give their full support to council officials, right or wrong; others are sometimes at loggerheads with officials. See *The New Local Authorities, Management and Structure*, HMSO, 1972.

B9   Bimodal distribution

A frequency distribution (F9) with two modes (M15).

B10   Binary numbers

Decimal numbers consist of the ten numbers 0 to 9; binary numbers, of the two numbers 0 and 1. In binary, then, 10 equals 2 of the decimal notation, and is described in terms of the decimal system as being of the *base 2;* the decimal system is described in its own terms as being of base 10. The following table of equivalents is a useful and revealing guide to conversion.

Binary	Decimal
1	1
10	2
100	4
1 000	8
10 000	16
100 000	32

By addition:

Binary	Decimal
11	3
101	5
110	6
111	7
1 001	9
1 010	10
1 011	11

It will be seen that, as the binary numbers shown proceed from 1 upwards in powers of 10 (binary), the corresponding decimal numbers proceed in powers of 2 (decimal), ie 2^0, 2^1, 2^2, 2^3, ...

Computers use the binary system internally for technical reasons. Print-outs are invariably stated in decimal numbers.

B11   Binominal distribution

A frequency distribution (F9) based on the expansion of $(x + y)^n$, where x and y represent the chance of each of two alternative events. If the probability of, say, success is 90%, leaving a 10% probability for failure then $x = 0.9$ and $y = 0.1$, and $x + y = 1$. The probability of two successes in two events is x^2, ie $0.9^2 = 0.81$, that of two failures, y^2, ie $0.1^2 = 0.01$, and that of one success and one failure is $xy + yx = 2xy$, ie $2(0.9)(0.1) = 0.18$. The sum of these three probabilities, $0.81 + 0.01 + 0.18$, is unity. The three probabilities are the terms of the expansion of $(0.9 + 0.1)^2$ ie

$$0.9^2 + 2(0.9 \times 0.1) + 0.1^2$$

With three events, there are four probabilities, namely:

$$0.9^3 + 3(0.9^2 \times 0.1) + 3(0.9 \times 0.1^2) + 0.1^3$$
$$= 0.729 + 0.243 + 0.027 + 0.001$$

which means that in three events the probability of

Three successes is	0.729
Two successes and one failure is	0.243
One success and two failures is	0.027
Three failures is	0.001
Total covering all possible results	1.000

If the number of events is the number of items in a sample drawn at random from a batch of a product, and successes are acceptances and failures are rejects, the same kind of reasoning applies. See addition law (A5), multiplication law (M22) and Pascal's triangle (P6). For Poisson distribution, see P19.

Application. Quality control.

B12   Bivariate distribution
A term applied to two series of figures, one for each of two related factors shown side by side. The two series of figures set side by side as plotted in Figure B5.1 provide an example. Often used by statisticians as a preliminary canter to show whether two series are significantly correlated (C25).

B13   Branding
A subject that has exercised the minds of economists for many years on account of the effect on competition. Its introduction results in a change from perfect competition (P12) to a monopoly situation, and, therefore, a fall in the price elasticity of demand (D7) from

infinity to a level that enables the producer to exercise a measure of control over his selling price (R6). Where competition is perfect, as it is in cattle markets, metal markets and other wholesale markets in staple commodities, price is determined entirely by the market forces of supply and demand. A condition of effective branding is that consumers must be persuaded that the brand is desirably different in some respect from the product in general.

Branding has proved to be so profitable to manufacturers and food processors that some growers of produce have attempted to exploit it by packaging their products in branded transparent bags. Whether the High Street has yet seen the ultimate in branding remains to be seen. Cheese with its many named varieties — not to be confused with brands — appears to be unbrandable. Would a brand of, say, Cheddar be acceptable to the epicure? If it is different from Cheddar in flavour or texture, a necessary condition of effective branding, can it fairly be called Cheddar? Yet there are now brands of Cheddar and other varieties of cheese to be found in the grocers' shops. Perhaps this is the ultimate in branding. Perhaps it even goes beyond the ultimate.

There is a statistical test of the value of the monopoly which a supplier holds in his brands. It is to determine the implied elasticity of demand (I2) of each brand and to decide whether it is reasonable in the light of experience. In the course of time it may become possible to exploit the monopoly fully by applying the principles of rational price fixing (R6) to the brand. Where weighing and packaging machinery is provided, an investment appraisal may be worthwhile.

Applications. Pricing policy.

B14 Break-even analysis

A technique to determine the rate of sales in quantity so that the rate of money sales can be equated to the rate of costs; usually expressed in terms of annual rates. There are two approaches: one, the linear, in which it is assumed that the selling price of the product remains constant, differences in the rate of sales being due to other factors; the other, the curvilinear, in which it is assumed that differences in the rate of sales are due to differences in the selling price.

Linear approach. Figure B14.1 demonstrates this. The cost

graph, *T,* is a straight line intersecting the vertical axis at a point, *F,* which is the measure of the fixed annual costs (F8), and *S* is the sales graph, which is also a straight line since the price is held constant for all rates of sales. The point of intersection of the two graphs is the break-even point, Q_v on the horizontal axis being the break-even quantity, S_v on the vertical axis, the break-even money sales. Figure B14.1 shows how the problem is usually depicted in the literature.

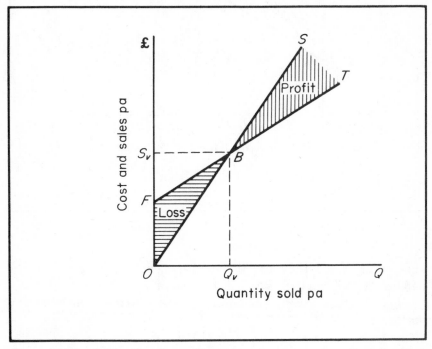

Figure B14.1   A linear break-even diagram

Curvilinear approach. Figure B14.2 provides a diagrammatic demonstration of the curvilinear analysis. The graph of total cost, *T,* remains a straight line, as in B14.1, but the sales graph, *S,* becomes a curve, concave downwards as it rises to the right (see S1). Here, there are two break-even points, namely *B*1 for a relatively high price, and B_2 for a relatively low price. Commercial undertakings are less interested in the break-even points in this kind of analysis than in the optimum price which the diagram indicates. The optimum price is the price that yields the maximum net profit, the measure of which is represented by the vertical distance between the cost graph and the

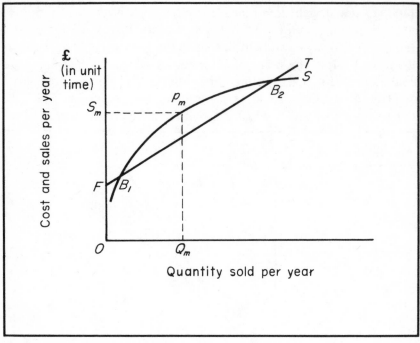

Figure B14.2   Maximizing net profit

sales curve where they are the furthest apart, ie where they are parallel. See P17 and 40, R6, M1, 3, 4 and 6.

Applications. Profit planning, price fixing, make-or-buy decisions.

B15   Budget

A table mostly expressed in terms of money values of a company's hopes, aspirations and expectations in the forthcoming period, usually the company's financial year. It covers sales, annual costs, capital expenditure and capital receipts, and may go into considerable detail about sales proceeds from different brands, costs of different materials, wages and salaries bills, and miscellaneous receipts and expenses on current and capital account. The company budget is often regarded as a statement of policy, but since its main basis is often a set of forecasts of sales, annual costs and capital expenditure, it is rather more: it is as much a statement of expectations as of policy.

B16   Budgeting

The process of making a company budget, and using the budget to exercise control over the organization as a whole and the various sections of it, by comparing actual results with budgeted results month by month or quarter by quarter. The use of the budget for control purposes is usually spoken of as *budgeting control.*

B17   Business ratios

Sometimes called efficiency ratios since their main purpose is to pinpoint areas of inefficiency in the organization. They are also used for target setting. The prime ratio is said to be profit to capital employed. Nowadays, the tendency is to build up ratio systems in which there may be a key ratio equal to the sum of the underlying ratios. For these, the same denominator is used throughout the system, such as capital employed, equity capital, turnover, net assets or value added. There is scarcely any limit to the numerators that can be used, all expressed in terms of money. Where gross-revenue turnover is used as the denominator, an all-embracing system can be built up, showing the proportions of gross revenue derived from the sales of different products, and the proportions spent on wages, salaries, materials, transport, etc. and the proportion going to profit. In such an all-embracing system, there may not be a key ratio, but there may be a number of subkey ratios. The sum of all the ratios may be one (or 100 where the ratios take the form of percentages). Provided the time unit is short, say, a week or a month, and the basic figures are seasonally corrected, business ratios provide a very effective and sensitive means of detecting areas of creeping inefficiency in the organization. They may be divided into three categories: those which the company hopes to see rising, eg profit to turnover, those which it hopes to see falling, eg cost to turnover, and those which it hopes to see remaining constant at an optimum, eg the value of stocks to turnover. Systems of business ratios are employed by the Centre for Interfirm Comparison Ltd (see I12), a subsidiary of the British Institute of Management, as they are so revealing of the relative efficiency of each subscribing company, without giving away the identity of any company, or any confidential information.

C1   Cake diagram
Same as pie chart (p 15).

C2   Capital employed
A term variously used to denote one of a number of things, generally one of the following:

(1) *Total capital:* the total of share capital issued and paid-up, loan capital and reserves.
(2) *Shareholders' capital:* the total of paid-up ordinary and preference share capital and reserves.
(3) *Equity capital:* the total of paid-up ordinary share capital and reserves.

C3   Capital method
The process of adding to, or deducting from the capital outlay (C4) on a new work, the present value (P22) of any expenditure or receipts of a capital nature accruing after the work has been brought into use. See revenue method (R16).
    Applications.   Investment appraisal, efficiency auditing.

C4   Capital outlay
Expenditure on providing additional or replacement fixed assets and additions to stocks of materials. It includes expenditure on the purchase of leasehold and freehold land and buildings and the goodwill of a business taken over, but excludes expenditure on the repair and maintenance of fixed assets other than the cost of major replacement parts.

C5 Capital works, classes of

Apart from works designed to reduce external pollution, internal health hazards and to improve the local amenities, there are, theoretically, 18 classes of capital work that need to be recognized. The first division is into two categories made by reference to source of yield: (1) by additional gross revenue, indicating expansion; (2) reduction in annual costs, indicating saving. The second division is made by reference to the effect on the company's capital stock, of which there are three classes: (1) additions or betterment, ie capital development; (2) replacement or renewals; (3) transfers and displacements under reorganization schemes. The third, and final division is made by reference to the effect on the company's cost structure, of which there are three classes: (1) a change in fixed costs (F8) but not in marginal costs; (2) a change in marginal costs but not in fixed costs; (3) a change in both.

The six yield-capital stock classes are:

Expansion	*Saving*
Development	Development
Renewal	Renewal
Reorganization	Reorganization

and each one of these can be subdivided into the three cost-structure classes, making 18 classes in all. Needless to say, not all commercial capital works fit snugly into one or other of these classes. Some large-scale reorganization works, for instance, involve either development or renewal or both, and, thanks to technology, some saving renewal works may also be expansionary in the sense that they result in an increase in capacity, as well as in a saving.

Applications. Investment appraisal, efficiency auditing, growth policy.

C6 Cash flow

Has four distinct meanings in the literature, namely

1 Gross revenue.
2 Net revenue net of depreciation.
3 Net revenue gross of depreciation.
4 The flows in both directions of actual money (currency, cheques and so forth).

Accountants usually mean the second of these when they speak of cash flow; financial journalists mean the third. The last meaning has come recently into use in the phrase *cash-flow reporting* or *cash-flow accounting,* under which, actual movements of money on capital and revenue account combined are reported, instead of the totals of debits and credits. The system was devised during a period of credit restriction, when many companies experienced difficulty in keeping liquid enough to continue in business. Apart from this aspect, the system appears to have nothing to recommend it.

C7   Causal relationship
Statistical analysis can go a long way towards establishing the existence of a causal relationship between two measurable factors. It cannot establish the nature of any causal relationship, nor can it be used for proving that any two factors are not causally related.

So far as the nature is concerned, statistical analysis may show that factors *A* and *B* are causally related; it cannot show that *A* causes *B* or that *B* causes *A,* or indeed, that either one causes the other, since the relationship shown to exist between *A* and *B* may be the effect of factor *C,* ie it may be that *A* and *B* are joint effects of *C.*

As to factors that, on the evidence of the statistics, seem not to be causally related, *B* and *C* may be joint causes of *A,* but *C* may behave so erratically relative to *B* that the statistics of *A* and *B* fail to reveal any causal relationship. Table C7.1 contains a hypothetical example of a problem in three variables, in which *A* is dependent on *B* and *C,* and in which the statistics for *A* are built up from the equation $A = 4B + 10C.$

Table C7.1   Causal relationship with three variables.

A	B	C
158	2	15
148	5	12
120	10	8
134	11	9
132	8	10

A comparison by inspection of the two series representing *A* and *B,* or for that matter, *A* and *C,* suggests little causal relationship between the two factors. Yet if we could compare A with B and C together, we would find the statistical relationship to be 100%. There is a statistical measure of the relationship between two or more statistical series: it is called the coefficient of correlation (C25).

Apart from the kind of problem exemplified in Table C7.1, there is another which statistical analysis is at least unnecessary for solving. If B were what is known as a necessary and sufficient condition to A, then A never happens when B does not happen, and whenever B happens, A also happens. A statistical inquiry to prove that A and B are causally related in such circumstances would be entirely superfluous. Since lung cancer occurs in people who have never smoked, cigarette smoking is not a necessary condition to lung cancer, and since many lifelong smokers of cigarettes never contract lung cancer, cigarette smoking is not a sufficient condition. In the absence of biochemical evidence, a statistical inquiry — such as that of Drs Doll and Bradford Hill in 1950 — could be useful in establishing a causal relationship between lung cancer and cigarette smoking. But the investigators' conclusion from the evidence of the existence of a causal relationship, that the latter caused the former, was not strictly a scientific one. It was, at the time, no more than a probably valid conclusion, for there were other possible explanations, such that a propensity to lung cancer causes a craving for tobacco smoke. In any event, that cigarette smoking is neither a necessary condition nor a sufficient one, goes to show that other factors play a part.

An understanding of the logic of causal relationships is of use in statistical research in all fields.

C8   Census of distribution

A census of wholesale and retail trades in Britain was first taken for the year 1950. This was a complete census; since then, several more censuses of distribution have been taken, some of them sample surveys. The censuses include certain retail service trades, and the reports published by HMSO give a considerable amount of detail. Some data are also given in the *Annual Abstract of Statistics*, HMSO.

C9   Census of production

The first census of production of manufactured goods related to the year 1907. Many more, mostly sample surveys, have been taken since. Full censuses call for a great amount of detail, most of which is published in the report on each detailed census (HMSO), summary information being given in the *Annual Abstract of Statistics,* HMSO.

C10   Central tendency
Averages, especially the mode (M15), are said to be measures of central tendency.

C11   Chain-base index number
A type of index number (I7), which changes its base (=100) and its pattern of weights from one period or from one date to the next. Both Laspeyres's formula (L4) and Paasche's formula (P1) can easily be converted to the chain base. Laspeyres's standard fixed-base price index number in its more practicable form is

$$P_{n.\,0} = \frac{\Sigma(q_0 p_0 \frac{p_n}{p_0})}{\Sigma(q_0 p_0)} \tag{11.1}$$

where P is the price index, Σ the sign of summation, q the quantity, p the price, and subscripts n and 0, represent the given and base periods, respectively, so that q_0, p_0 are the quantity and price in the base periods, and $P_{n.\,0}$ is the price index in period n, based on period $0 = 1.00$. The corresponding chain-base formula is

$$P_{n.0} = P_{(n-1).0} \frac{\Sigma\left(q_{n-1} p_{n-1} \frac{p_n}{p_{n-1}}\right)}{\Sigma(q_{n-1} p_{n-1})} \tag{11.2}$$

which gives the index number to the original base = 1.00. It will be seen that the weights used are those of the previous period. Reference to the original base is a convenience, and is made, as (11.2) shows, by multiplying $P_{n.(n-1)}$ by $P_{(n-1).0}$, so that the index to the original base for the given period, n, is the product of the index numbers each derived from consecutive periods:

$$P_{n.\,0} = P_{1.0} \cdot P_{2.1} \cdot P_{3.2} \cdots P_{(n-1).(n-2)} P_{n.(n-1)}$$

Chain-base formulae are useful where items tend to change in design and specification so frequently that, in the course of time, a representative sample with comparable data for both the base (original) period and the given period cannot be assembled. Consumer durables, for which styling and design are important selling points, are particularly prone to change in this way. Then the chain index provides a solution: it should nearly always be possible to draw a representative sample with comparable data for two consecutive periods. There is one fairly serious objection to the chain-base index. Where price changes tend to take place more often with changes in design or specification than otherwise, the selection in each period of a new sample — which would by definition exclude items whose specification has changed since the previous period — gives a misleading result; it would amount to throwing out the baby with the bathwater.

C12   Chi-squared test

A test to indicate whether a set of observed frequencies differs from the expected frequencies, by chance or otherwise. *Chi* is pronounced *kye,* and refers to the Greek lower-case letter χ, equivalent in sound to the English *ch.* In the literature, the test is usually denoted as the χ^2 *test.* The formula in its simplest form is

$$\chi^2 = \Sigma \left(\frac{(A\text{-}F)^2}{F} \right)$$

where A represents the actual frequencies, and F the expected frequencies. It will be seen that, where the actual frequencies conform entirely to the expected, the value of χ^2 would be zero. The further this value departs from zero, the more likely it is that there is some element other than chance present. Mathematical statisticians have compiled tables of the values of χ^2 attributable to chance for ranges of probabilities and of degrees of freedom. A probability of 5% would mean that the odds are 20 to 1 against the figure shown in the cell of the table being exceeded other than by chance. Whether a 5% probability is good enough, depends on the circumstances. An element of personal subjective judgement must necessarily enter into the decision but, at least, we can ascertain the probability.

Suppose it is suspected that one of a set of dice is loaded. The χ^2 test can be used for ascertaining the degree of probability that it is. The suspected die is thrown 60 times, the actual frequencies being as shown in Table C12.1, which also gives the expected frequencies, ie 10 times for each face, and the calculations to give the value of χ^2.

Table C12.1 Calculating a simple χ^2 distribution with a die thrown 60 times.

							Total
Face, number of pips	1	2	3	4	5	6	
Expected frequencies, F	10	10	10	10	10	10	60
Actual frequencies, A	4	7	10	9	12	18	60
$A - F$	−6	−3	0	−1	+2	+8	0
$(A - F)^2$	36	9	0	1	4	64	114
$(A - F)^2/F$	3.6	0.9	0	0.1	0.4	6.4	11.4

There are six faces and, therefore, five degrees of freedom. The value of χ^2 turns out to be 11.4. Reference to the table shows that the probability is about 5% that chance accounts for the differences between the actual and expected frequencies. This may not be good enough to justify accusing the owner of the set of dice of cheating, but it would be good enough to advise having the suspected die examined to ascertain the location of its centre of gravity.

Applications. Sampling, quality control, testing frequency distributions.

C13 Classification

'The first efforts in the pursuit of knowledge then, must be directed to the business of Classification. Perhaps it will be found in the sequel, that Classification is not only the beginning, but the culmination and the end, of human knowledge.' (F. Bowen, *A Treatise on Logic*, Harvard, 1866.)

A sound classification of the items in the field – the species of the genus – is as fundamental in the collection and collation of statistics as it is in other spheres. There are five attributes of a sound classification:

1   Its orders, divisions and classes are all mutually exclusive.
2   It is exhaustive of the genus.
3   Its orders, divisions, and classes are relevant to its purpose.
4   Its divisions are practicable and realistic.
5   Written headings are concise and clear.

Mutual exclusiveness is achieved by dividing orders, divisions and classes, each in turn, by reference to a single characteristic such as colour, material, size or function.

Exhaustiveness in detail is important in a classification of factory stores. In purely general statistical classifications, such as the Export List, exhaustiveness is often achieved by the use of residual headings: 'Other', 'Other types', etc. the so-called *rag-bag* headings (R 1).

Relevance is largely a matter of common sense and a due regard to the purposes of the classification. A division of machine tools by reference to colour would generally be irrelevant, though it may be useful to paint manufacturers.

Practicability is an attribute that is frequently overlooked. It can be said that the practicable dividing characteristics are internal to the individual items themselves. End use is an example of an external, and generally impracticable characteristic. Immediate function is internal and always practicable. A washing machine is internally designed for washing clothes, but not for any particular end use such as washing towels in an hotel.

Conciseness and clarity provide a big subject. It should be kept in mind, when compiling a classification, that the headings are not intended to give definitions. Anything more than necessary to distinguish items one from another would be superfluous. Clarity can be achieved only by a careful choice of word and phrase. Some electronic-computer failures are due to weaknesses in the classifications underlying the numerical coding system.

C14   Coefficient of variation
A standardized comparative measure of dispersion (D19). It is equal to the standard deviation (S24) of a series of figures, divided by the arithmetic mean (A14) of the series. The standard deviation provides an absolute measure of the dispersion of a series in the same terms as the series itself — tons, yards, £s, number, etc. The coefficient provides a means of comparing the dispersions in two or more series.

C15   Combinations
A combination consists of a selection of any number of items from a group of items. Six combinations, each of two things, can be selected from a group of four things. From an apple (*a*), a pear (*p*), an orange (*o*) and a lemon (*l*), taken two at a time, the six combinations are

ap, ao, al, po, pl, ol

The formula for determining the number of combinations that can be obtained from n objects taken r at a time is

$$\frac{n!}{r!(n-r)!}$$

For the example of the fruit, we therefore have

$$\frac{2 \times 3 \times 4}{2 \times 2} = \frac{24}{4} = 6$$

A set of least-squares (L7) equations (based on mean deviations) for four variables consists of four. But only three at a time can be used for deriving the three coefficients. The number of solutions for each coefficient is equal to the number of combinations of three equations, ie

$$\frac{2 \times 3 \times 4}{2 \times 3} = \frac{24}{6} = 4$$

so that there are four solutions for each coefficient.

It is worth noting that the factorials in the denominator and numerator cancel out to some extent. In football-pool language, combinations are wrongly referred to as *permutations*, for which see P13. The number of combinations of eight of 10 selected games, for the treble-chance pool is

$$\frac{2 \times 3 \times 4 \times 5 \times 6 \times 7 \times 8 \times 9 \times 10}{2 \times 3 \times 4 \times 5 \times 6 \times 7 \times 8 \times 2}$$

which, after cancelling out, gives $90/2 = 45$. Out of a selection of 11 games, it is

$$45 \times \frac{11}{3} = 165$$

and of 12 games,

$$165 \times \frac{12}{4} = 495$$

Applications. Of general use to statisticians.

C16   Compounding
The reverse of discounting (D16); the process of determining the amount of a sum of money invested at fixed compound interest for a number of years, or the amount of a sum of money invested each year at fixed compound interest. For formulae, see A8 and 9.

Applications. Investment appraisal, efficiency auditing, pension-fund calculations.

C17   Confidence limits
Fiducial limits (F2).

C18   Constant
A mathematical term for a figure that remains constant or is assumed by some hypothesis to remain constant. There are the universal mathematical constants such as *pi* and *epsilon,* conversion factors such as knots to feet per second, and the so-called physical constants such as the velocity of light. Special constants of interest to management statisticians include marginal costs (M4), the elasticities of demand with respect to price and to other factors (D7) and other regression coefficients (R8).

C19   Consumers' expenditure
Probably the most important of environmental statistics for business management.

Sources. National Income and Expenditure, annually (September), HMSO; *Economic Trends,* monthly, HMSO, gives quarterly figures; an index of retail sales is published monthly in the *Monthly Digest of Statistics;* also the Census of Distribution reports (C8).

C20   Contingency
A chance occurrence; an event incident to another. Estimates of current and capital expenditure often include an item for contingencies, necessarily unspecified, though the compilers of estimates may sometimes have in mind some particular event or events that may or may not occur.

C21   Continuous variation
Items expressed in terms eg of weight, volume, length and money value, are continuous in their variation. Items expressed in terms of whole numbers are not continuous and are said to be discrete in their variation (D17). Quantity figures of oranges sold by weight are continuous, those of oranges sold by number of units or crates are discrete.

C22   Contribution
A management-accounting term meaning the difference between the

price of an article and the marginal cost. The difference is the article's contribution to overheads and profits. Also used in the aggregative sense of the difference between the sales proceeds and the variable cost of sales (C31). See *P/V* ratio (P36).

Applications. Profit planning, break-even analysis, net-revenue analysis.

C23 Control chart
Used of internal time charts in general, and of quality control charts (Q3) in particular. See group internal control chart (G10).

C24 Coordinates
The dots in a scatter diagram (S9); regarded as the points of intersection of straight lines that might be extended horizontally from the vertical axis, and vertically from the horizontal axis.

C25 Correlation
A concept which has two connotations: (1) the true correlation between two factors; (2) the statistical correlation. Usually it is possible to theorize only about the former; the analyst, the pragmatist and the rest have to be content with the latter. The odd thing is that where two factors can be shown by general reasoning to be causally related (see C7), then other things being equal, there is usually 100% correlation between them, ie the true correlation is 100%. If this is valid, it follows that any measure of statistical correlation is strictly an inverse measure of the forces exerted by other factors in the relationship, and not at all an indication of the true correlation.

There are two measures of statistical correlation: (1) the coefficient of simple correlation (existing between two factors); (2) the coefficient of multiple correlation (existing between a dependent factor and two or more independent factors). For simple correlation the formula is

$$r = \frac{\Sigma(xy)}{\sqrt{(\Sigma x^2 . \Sigma y^2)}}$$

where r is the coefficient, Σ the sign of summation, and x and y the deviations from the means of factors X and Y, respectively. For multiple correlation, the formula is

$$R = \sqrt{\left(1 - \frac{S^2}{\sigma_1}\right)}$$

where R is the coefficient of multiple correlation, S the standard error of estimate (S25) of the regression (R7) of the dependent variable X_1 on the independent variables X_2, X_3 ..., and σ_1 the standard deviation (S24) of X_1.

With simple correlation the coefficient ranges from -1.0 for perfect negative correlation (N2) to $+1.0$, usually written without the plus sign, for perfect positive correlation (P21). With multiple correlation, the coefficient ranges from 0 to 1.0, there being no such statistical entity as negative multiple correlation.

Table C25.1   The coefficient of simple correlation.

Reduced values

	A	B	A−120 (= Y)	B−2 (= X)	Y^2	X^2	XY
	158	2	38	0	1 444	0	0
	148	5	28	3	784	9	84
	120	10	0	8	0	64	0
	134	11	14	9	196	81	126
	132	8	12	6	144	36	72
Total	692	36	92	26	2 568	190	282
Average	138.4	7.2	18.4	5.2			
Correction sum, subtract					1 692.8	135.2	478.4
Sums of squares and products of					875.2	54.8	−196.4
mean deviations					Σy^2	Σx^2	$\Sigma(xy)$

In the example of Table C7.1, we see by inspection that there is not much simple statistical correlation between A and B or between A and C. Consider the correlation between A and B, for instance. Table C25.1 sets out the necessary calculations, using y for the mean deviations of A, and x for those of B. It will be observed that two short cuts are employed, both useful when computer time is not available, with the object of saving arithmetic and labour time. One is to deduct from each series of figures the lowest figure in the series; the other is to square and multiply the figures themselves, instead of the deviations from their means, and make a simple correction to the totals to obtain the sums of squares and products of the mean deviations. The correction sums are the squares or products of the sums of the reduced series divided by the number of sets of observations, thus

$$92^2/5 \quad = \quad 1\ 692.8$$
$$26^2/5 \quad = \quad 135.2$$
$$\frac{92 \times 26}{5} \quad = \quad 478.4$$

The coefficient of simple correlation based on the above formula is

$$r = \frac{-196.4}{\sqrt{(875.2 \times 54.8)}} = \frac{196.4}{219.0}$$

$$= -0.897$$

There is, therefore, a fairly high degree of negative correlation between the two series, A and B. However, this does not take into account the effect on A of factor C. The example of Table C7.1 is based on the equation $A = 4B + 10C$, so that the deviations from the actual values of A in the table, of any estimate of A derived from this equation, are all zero. The standard error of estimate is therefore also zero, and the coefficient of multiple correlation is 1.00:

$$R\sqrt{(1-\frac{0}{\sigma_1})} = 1.00$$

The fact that the sign of $4B$ in the equation is positive shows that A and B are positively, not negatively, correlated. We can go one step further. The effect on A of factor C can be eliminated by deducting the values of $10C$ from the corresponding actual values of A. By correlating the values of A thus adjusted with the series for B, we should obtain a much more accurate value of r. Table C25.2 gives the necessary calculations. As the adjusted values of A are now necessarily all equal to $4B$, the correlation existing between them is positive and perfect, and $r = +1.00$. Table C25.3 shows this to be correct. The figure 1.00 is called the coefficient of part correlation (P4).

Admittedly this is an extreme example. It is not often that the statistical correlation between two factors would turn out to be negative where the true correlation is positive. But the example goes to show that the coefficient of correlation is not a reliable indication of the true correlation. In real life, there is rarely — if ever — any possibility of eliminating all other factors, except in controlled laboratory experiments where analysis is carried out by mathematics rather than statistics. Statistical methods are necessary only where unknown factors or non-measurable factors affect a relationship.

Table C25.2   Eliminating factor C from A.

A	C	10C	Adjusted A (= A − 10C)
158	15.0	150	8
148	12.8	128	20
120	8.0	80	40
134	9.0	90	44
132	10.0	100	32

A multiple-regression analysis (M23) provides a measure, subject to error, of the effect of each independent variable on the dependent variable, and so the effect of each can be eliminated from the series representing the dependent variable. If the effect of all but one is eliminated, then the coefficient of correlation between the one remaining and the adjusted series of the dependent variable approaches more closely to the true correlation than that between the one factor and the unadjusted series. Even so, owing to the presence of unknown and unmeasurable factors, multiple regression can never be an accurate measure of the true correlation. When all independent variables that are taken into account in the regression analysis are eliminated, together with the unattached constant, there remains a residue for each observed value of the dependent variable. These residues form the basis of the standard error of estimate (S25).

Note that the two formulae given above apply to large numbers of sets of observations, generally referred to as large samples. Where correlation analysis is applied to small samples, the formulae are somewhat different, and the resulting coefficients are said to be corrected or adjusted for smallness of sample. The symbols are usually distinguished in some way, for instance by a horizontal bar surmounting them, ie $\bar{r}$ and $\bar{R}$. The two formulae are as follows:

$$\bar{r} = \sqrt{\left\{1 - (1-r^2)\left(\frac{n-1}{n-2}\right)\right\}}$$

$$\bar{R} = \sqrt{\left\{1 - \left(\frac{\bar{S}^2}{\sigma_1^{\,2}}\right)\left(\frac{n-1}{n}\right)\right\}}$$

where n is the number of sets of observations, and the horizontal bar over S indicates that the standard error of estimate is corrected for smallness of sample.

Table C25.3 A case of perfect positive correlation.

Adjusted A	Reduced adjusted A (= Y)	Reduced B (= X)	Y^2	XY	X^2
8	0	0	0	0	
20	12	3	144	36	From
40	32	8	1 024	256	Table
44	36	9	1 296	324	C25.1
32	24	6	578	144	

Total 144	104	26	3 042	760	
Average	20.8	5.2			
Subtract correction sums			2 163.2	540.8	
			878.8	219.2	54.8
			Σy^2	$\Sigma(xy)$	Σx^2

$$r = \frac{219.2}{(878.8 \times 54.8)} = \frac{219.2}{219.4}$$

$$= 1.00$$

What constitutes a small sample — and what, a large one — seems to be a matter of subjective judgement. The result is that some statisticians tend to use the $\bar{r}$ and $\bar{R}$ formulae on all occasions, no matter how large the sample may be. There is something to be said for this. The greater the value of n, the nearer that $(n-1)/(n-2)$ and $(n-1)/n$ approach to unity and so make the correction for smallness of sample that much smaller.

Applications. In statistical research generally, in helping to get the 'feel' of statistics prior to or during analysis.

C26 Cost analysis

An analysis of current costs generally designed to determine the marginal cost (M4) of a product brand, or to separate the total variable cost from the total fixed cost of running a factory or establishment. The method of finite differences (F3) applied to two or three sets of observations provides a useful means of carrying out the analysis. Care needs to be taken in such exercises to ensure comparability of basic data in respect of wage rates, materials prices, etc. (C28).

Applications. Profit planning, break-even analysis, product costing.

C27 Cost benefit

A term used variously, and vaguely. Perhaps its most useful definition is the cost (input) used as a measure of the output of non-trading services provided by central and local government and charities.

C28 Cost function

Cost is a measure of input. Whether input is a function of output, or output a function of input, is the kind of question that runs all through management statistics. When the method of least squares (L7) is applied, it becomes a problem as it does with mathematical model building (M8 and M16).

However, the functional relationship between cost (or input) and output usually takes the form of a cost function, ie cost is regarded as the dependent variable, and output as the independent variable. For a single homogeneous product, the algebraic function or model may be written:

$$T = aQ + F \qquad (28.1)$$

where T is the total annual cost, a the marginal cost, Q the annual output in terms of quantity, and F the annual fixed cost. This is a linear equation, whose graph is a straight line dissecting the vertical axis at F and rising to the right, as in Figures B14.1 and B14.2. It applies only to rates of output within the normal capacity of the plant (P17). When the normal capacity of the plant is exceeded, the special measures taken cause an increase in the marginal cost, and the graph curls upward from the straight line.

A cost function for two or more products may take the form either of the above, when Q becomes an index number (I7) of the firm's production, or of the following:

$$T = a_1 Q_1 + a_2 Q_2 + \cdots + a_n Q_n + F \qquad (28.2)$$

where $Q_1, Q_2, \ldots Q_n$ represent the firm's several products, and $a_1, a_2, \ldots a_n$ are the respective marginal costs. Equation (28.2) is of little use as a mathematical model for a regression analysis, except where the several products are limited to four or five. There would scarcely ever be enough degrees of freedom (D6) to permit more. In any event it is important, when deriving a cost equation of this kind from basic statistics, to see that extraneous factors in cost, such as wage rates and materials prices, are rendered constant.

Applications. Break-even analysis, profit planning, price fixing, net-revenue analysis.

C29 Cost of capital
The interest payable on loan capital.

C30 Cost-of-living index
An index number (I7) of retail prices made by the Department of Employment, and published monthly. Cost-of-living-index is the old name; it is now called the *retail-price index* or the *index of retail prices.* As the index is used extensively for a wide range of purposes, the Department goes to a great deal of trouble in compiling it. The formula used is Laspeyres's index (L4). Its pattern of weights is based on family-budget inquiries, which are conducted from time to time – the most recent one at the time of writing being that of 1969 – and the index is rebased and reweighted accordingly. It has been said that this makes the index more of the Paasche type (P1) than of the Laspeyres type, but this is not so. A Paasche index is currently weighted; the retail-price index is never currently weighted: the pattern of base-period weights is merely brought up to date from time to time, a process that makes it what might be called a hybrid chain-base index (C11). The present index includes rent, rates, other housing costs, public transport, travel, other retail services, fuel and light, meals out and holidays, as well as food, clothing, durable household goods and other goods.

Applications. As a deflator to convert money values to real values and in wages negotiations.

C31 Cost of sales
The production and distribution costs of the sales made in any period. In the manufacturing and agricultural industries there is a time lag (L3) between input and output, so that the costs incurred in any year cannot be related to the sales made in that year. The cost of sales eliminates the time lag.
Applications. Net-revenue analysis, rational pricing, break-even analysis.

C32 Cost schedule
A table setting out the total annual cost for each of a range of annual outputs under conditions of constant wage rates, materials prices, etc.; especially useful where rates of output in excess of.the normal capacity of the plant (P17) are involved. See net-revenue schedule (N6), also S32.
Applications. Statistical presentation, net-revenue analysis, price fixing.

C33 Cost-structure ratio
The ratio of a company's total annual variable cost (V5) to the total annual cost. The cost structure is said to be flexible where the ratio is relatively high, and inflexible where it is low. With the advance of technology, cost-structure ratios are tending to fall, which means that industry is finding it more and more difficult to reduce costs when demand is falling.
Applications. Investment analysis, investment appraisal, efficiency auditing.

C34 Cost variance
The difference between the actual cost and the standard cost. See analysis of variance (A10), also S23.

C35 Covariance
A measure of the variance between two series representing two factors. The formula is

$$\text{Covariance} = \frac{\Sigma(xy)}{n}$$

where Σ represents summation, x and y are the mean deviations of X and Y, and n is the number of sets of observations. The covariance between factors A and B in Table C25.1 is

$$- \frac{196.4}{5} = 39.3$$

which indicates that the two series are negatively correlated (N2), and the extent of the covariation. The covariance between A and B in Table B25.3 is

$$\frac{219.2}{5} = 43.8$$

which indicates positive correlation (P21) and a somewhat larger covariation.

Applications. Statistical research involving correlation analysis.

C36 Cross-classification

One of the simpler methods of multiple-regression analysis (M23) applicable to problems in up to five variables. It is, in practice, an adaptation of the simple-regression method of group averages (G9) to problems in three or more variables. Although a method of multiple--regression analysis in its own right, it appears to be used mainly in empirical model building (M16), for determining the type of mathematical model that best fits the statistics. Where the type of mathematical model is known to be linear, a two-tier system of cross-classification can be used for a regression analysis, but where the object of applying the method is to determine the appropriate type of model, a three-tier system is necessary.

In its original form, as demonstrated in the textbooks, cross-classification calls for the use of a three-dimensional working sheet. For a three-tier system, a problem in three variables provides for nine main cells. With an average of four sets of observations to the cell, the total number of observations required would be 36. For a problem in four variables, there would be 27 main cells, needing a minimum of 108 sets of observation.

A modern variant is to take each independent variable separately. For a three-tier system, there are three main cells only, and if a minimum of five sets of observations to a cell are considered sufficient, the total number of sets would be 15; this holds good for problems in any number of variables. The first step is to arrange the

Table C36.1 The modern variant of cross-classification.

		(Basic data and first analysis)				
	Weekly figures			Successive differences of averages		
Miles	Calls	Man hours		Miles	Man hours	a =
M	C	H		M^1	H^1	H^1/M^1
50	80	40				
76	90	44				
80	70	49				
84	35	48				
90	55	54				
Total 380	330	235				
Average 76	66	47				
106	70	60				
110	85	65				
112	50	64				
114	65	64				
118	70	67				
Total 560	340	320				
Average 112	68	64		36	17	0.4722
120	70	69				
121	50	68				
122	60	65				
124	85	72				
128	70	71				
Total 615	335	345				
Average 123	67	69		11	5	0.4545
Overall total 1 555	1 005	900				
Average 103.7	67.0	60.0		47	22	0.4681

sets of observations in order of magnitude of one of the more important independent variables; the second, to divide them into three groups of more or less equal size; the third, to average each variable in each group; fourthly, to see that the averages of the other independent variables are much the same down the columns; and finally, to apply the method of finite differences (F3) to the successive differences in the independent variable and the dependent, ie to divide the former into the latter.

If there are more than two independent variables, the data should then be rearranged in order of magnitude of another of them, and the process repeated. The ratios of the successive differences of the dependent to the independent variable are estimates of the regression coefficient of that independent variable. In a three-tier system there are two ratios for each; if they are much the same, the inference is that the appropriate mathematical model is linear, if not, a type of curvilinear model may be appropriate.

As cross-classification is of great value in empirical mathematical model building, a worked hypothetical example is given in Table C36.1. A transport undertaking is engaged in delivering parcels for local retailers, and for costing purposes, it needs to know the average time spent per vehicle mile, and per call. Man hours per week depend on mileage and calls made, and it is proposed as a first experiment to assume the following linear mathematical model to serve as the basic equation of a regression analysis:

$$H = aM + bC + K$$

where H represents man hours, M vehicle mileage, C the number of calls made, a and b are regression coefficients of M and C, and K is an unattached constant.

It will be seen that the figures of the average number of calls in the three groups are much the same as each other. It can therefore be said that the factor of calls is held constant, so that, apart from residual errors, the variations in the averages of H are attributable to variations in M.

The successive diferences, shown as M^1 and H^1 in Table C36.1, are derived from the averages: $112 - 76 = 36$, $123 - 112 = 11$, and so on. As their extraction has the effect of eliminating the constant K, the ratio of H^1 to M^1 provides an estimate of the value of a of the basic equation. However, the table uses a three-tier system, so that there are two ratios, one relating to the lower range of weekly mileages, and the other to the upper range. As there is little to choose between the two figures of 0.4722 and 0.4545, it can be taken that the linear assumption is, so far, proved to be empirically valid. If it is decided not to continue the test for linearity, the overall average ratio can be used. The overall averages of M^1 and H^1 shown at the foot of Table C36.1 are equal to the differences between the figures in the two extreme groups, and also to the sum of the two original averages of M^1 and H^1 :

$$123 - 76 = 36 + 11 = 47$$
$$69 - 47 = 17 + 5 = 22$$

This provides a check on the arithmetic.

There is no need to repeat the process demonstrated in Table C36.2 for determining estimates of the regression coefficient of the last remaining independent variable. It can be done in this example by eliminating the effect on the original H series of variations in the vehicle mileage, and applying finite differences to group averages. Table C36.2 shows how the process of elimination is carried out. The next step is to rearrange the corrected H series (symbolized by ^1H) and the C series into three groups in order of magnitude of the C series, as in Table C36.3 and to apply the method of finite differences, as in Table C36.1.

Table C36.2   Eliminating from H the effect of variations in M.

M	$0.468M$	H	^1H $(= H - 0.468M)$	C
50	23.4	40	16.6	80
76	35.6	44	8.4	90
80	37.4	49	11.6	70
84	39.3	48	8.7	35
90	42.1	54	11.9	55
106	49.6	60	10.4	70
110	51.5	65	13.5	85
112	52.4	64	11.6	50
114	53.4	64	10.6	65
118	55.2	67	11.8	70
120	56.2	69	12.8	70
121	56.6	68	11.4	50
122	57.1	65	7.9	60
124	58.0	72	14.0	85
128	59.9	71	11.1	70

It will be seen that the two estimates of b differ appreciably, one from the other, and so the question arises as to whether the linear assumption can be accepted as empirically valid, as it seemed to be from Table C36.1. This is the kind of problem that often crops up in

Table C36.3   Estimating the coefficient of C.

				Successive differences		
	C	^1H		C^1	$^1H^1$	$b = {}^1H^1/C^1$
	35	8.7				
	50	11.6				
	50	11.4				
	55	11.9				
	60	7.9				
Total	250	51.5				
Average	50	10.3				
	65	10.6				
	70	11.6				
	70	10.4				
	70	11.8				
	70	12.8				
Total	345	57.2				
Average	69	11.4		19	1.1	0.0579
	70	11.1				
	80	16.6				
	85	13.5				
	85	14.0				
	90	8.4				
Total	410	63.6				
Average	82	12.7		13	1.3	0.1000
Overall total	1 005	172.3				
Average	67	11.9		32	2.4	0.0750

practice, especially where the number of sets of observations per group is relatively small, as it is in this example. Small numbers fail to iron out from the averages the effect of random variations. If more sets of observations are not available to improve the representativeness of the averages, then an inspection of the basic data may throw some light on the problem. Consider, for instance, the first group in Table C36.3. The three figures for H in the middle of the group seem to be high relative to those for C. It may be found that the weeks to which they relate were characterized by fog, or man hours may have been inflated otherwise, eg by delays at the garage in the mornings.

It is worth mentioning here, that general reasoning tends to prove that, for man-hour analyses of this kind, the valid form of mathematical model is linear. See L1.

If it is decided to accept the linear assumption as valid and to complete the regression analysis by cross-classification, all that is required is the value of K in the basic equation. For this, the calculated regression coefficients are applied to the overall averages of the variables in Table C36.1:

$$H = aM + bC + K$$
$$\text{Therefore } K = H - aM - bC$$
$$\text{ie } K = 60.0 - 0.468 \times 103.7 - 0.075 \times 67$$
$$= 6.45$$

The complete regression equation is therefore
$$H = 0.468M + 0.075C + 6.45$$

A comparison with the solution by least squares is given under L7.

Interpretation. At 0.468 hours to the mile, the average speed between calls is 2.14 mph. The average time taken per call is 0.075 hours, ie 4½ minutes, and the average time taken by the driver in clocking on and off, taking meals, preparing the vehicle for the road, etc. is 6.45 hours a week.

Applications. Mainly in empirical testing for linearity in model building. But, as a method of multiple-regression analysis, cross-classification can be applied to the analysis of market demand, cost analysis, labour-time analysis (as demonstrated above) and in determining the scale of production and learning laws.

C37 Cross-section series

A term applied to any statistical series that relates to different things or different places at the same time, as distinct from a time series (T8) which relates to the same thing or the same place at different times. If the man-hours example of Table C36.1 related to 15 vehicles in the same week, the three statistical series would be cross-section series; if to one vehicle over a period of 15 weeks, they would be time series.

Sometimes hybrid series (H5) are used for analytical purposes. If, for instance, the transport firm exemplified in Table C36.1 had only five vehicles in use, then the three statistical series might relate to each of the vehicle's operations in each of three weeks, in which case the statistics could be described as hybrid.

C38   Cumulative total

A total of totals, usually a moving total month by month or year by year. In a cumulative-frequency distribution it is the accumulated frequency up to, or not less than a given item, an example being:

Item	A	B	C	D	E	F
Frequency	1	3	5	10	9	7
Cumulatives:						
'up to'	1	4	9	19	28	35
'not less than'	35	34	31	26	16	7

The items $A,B,C,...$ may be read as numbers in a discrete series (D17), or ranges in a continuous series (C21). See F9 and M21.

C39   Current ratio

A business ratio: the total current assets to the total current liabilities. Companies try to maintain a ratio of at least unity, or even 1.5, to ensure their future liquidity.

C40   Cyclical variations

Long-term variations with a time cycle of two or more years, as distinct from seasonal variations (S11). The old trade cycle, as it existed before 1939, had a time cycle of 10-11 years. Since 1945 there has been little evidence of a trade cycle at all, although it has been said that there is now a 4-5 year cycle. A peak was reached, it seems, in 1951-2, and another in 1957. From a 1958 trough the rise was continuous up to 1971, apart from two short pauses in 1962 and 1967. Cyclical variations were at one time the forecaster's nightmare, but now, it seems, he can almost afford to ignore them and base his forecasts on the long-term trend as represented by the appropriate internal and environmental statistics. See S39.

d

D1   Decile
In a frequency distribution there are nine deciles which divide the series into 10 equal parts. The fifth decile is the median (M14), and the first is the tenth percentile (P11).

D2   Decision making
The process of ranking (R5) several possible mutually exclusive courses of action in order of merit in accordance with some criterion such as profitability, and choosing one of them.

D3   Decision tree
A statistical device for estimating the optimum size of order for perishable goods, or of a plant for making any kind of standard goods. As originally depicted, the decision tree consists of a number of main branches, each representing a size of order or plant and each with one or more secondary branches representing judgements of the financial consequences. The main branches grow out of a box labelled *D* (for decision) situated on the left of the picture and extending to the right, the secondary branches extending further to the right, and the whole tree lying prone as though it had just been felled.

   In its most logical and practicable form, the decision tree exploits the frequency distribution. Estimated daily sales of, say, meat pies are distributed by reference to the estimated number of days in every 100, for which each number of daily sales is made. Suppose the grocer insists on selling his pies across the counter on the day of delivery and disposing of any left-over as waste. From past daily records, he expects to be able to sell at least 22 pies a day, ranging upwards to more than 50 a day. He expects to sell either 22, 23 or 24 on one day of the 100; 25, 26 or 27 on two days of the 100; 28, 29 or 30 on three days; on to a mode (M15) of 40, 41 or 42 on 18 days, and finally, to 47 or more on 29 days of the 100. (See Table D3.1 for the complete distribution.) He has to purchase them from the local bakery in multiples of a dozen, and as he has never bought more than four dozen a day, he bulks upwards of 47 together, to begin with at any rate. His daily purchases impose a constraint on his daily and total sales.

Table D3.1   A tabular form of decision tree.

Daily purchase	Daily sales	Frequency	Sales proceeds at 3p each	Profit: (4) less cost	Total profit in period of 100 days	
	(No.)	(Days)	p	p	(3) x (5) £100	Total £
(1)	(2)	(3)	(4)	(5)	(6)	(7)
	47	29	141	45	13.05	
	44	16	132	36	5.76	
	41	18	123	27	4.86	
4 doz. at	38	15	114	18	2.70	
24p = 96p	35	11	105	9	0.99	
	32	5	96	0	0.00	
	29	3	87	− 9	−0.27	
	26	2	78	−18	−0.36	
	23	1	69	−27	−0.27	26.46
3 doz. at	35	89	105	33	29.37	
24p = 72p	32	5	96	24	1.20	
	29	3	87	15	0.45	
	26	2	78	6	0.12	
	23	1	69	− 3	−0.03	31.11
2 doz. at	23	100	69	21	21.00	21.00
24p = 48p						

What is the optimum number of dozens to purchase? He pays 24p a dozen to the baker, and sells at 3p each. Table D3.1 sets out the estimated financial results of each of the three decisions. If he buys four dozen a day, he will make an estimated gross profit of £26.46 in

100 days; if he buys three dozen, a profit of £31.11; and if 2 dozen, a profit of £21.00. On this evidence, the rational decision is to buy three dozen pies a day.

There is no reason why the grocer should not build up a frequency distribution pie by pie instead of grouping in threes. It would be more accurate in any event. The grouping has been used here to save space.

It should be kept in mind that the number of pies going to waste is no criterion. If it were, then the decision would fall to two dozen a day, or to reduce wastage to zero, to one dozen a day. Profit is the only valid commercial criterion — that, at any rate, is the underlying assumption of the decision tree, as it is of all other business techniques.

If the decision tree had shown that of the three sizes of daily purchase the most profitable one was four dozen, then to make sure it was the true optimum, the grocer would have to extend his frequency distribution, by guesswork to begin with, to daily sales in excess of 48. He could assume that the distribution was symmetrical, ie that the number of daily sales above the mode would decline at the same rate that they rose, up to the mode. This would give a distribution above the mode resembling that below.

Daily sales (number)	Frequency (days)
59	2
56	3
53	4
50	7
47	13

This accounts for the whole of the 29 days against daily sales of 47 in Table D3.1. The frequency polygon (C21, F9) for the example, as extended to a supposed daily purchase of five dozen pies, is given in Figure D3.1.

The decision tree is not to be confused with the economic ordering quantity (E6), which is concerned with stores control through ordering frequency.

Applications. Retail and wholesale distribution, the investment decision, ordering policy.

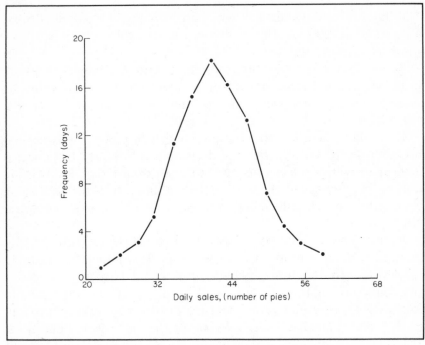

Figure D3.1   Frequency polygon of daily sales of meat pies

D4   Deduction

A priori reasoning; arguing from the general to the particular. Much of our everyday reasoning is deductive, eg 'Iron is a metal and therefore an element' is valid only if we accept the general proposition that all metals are elements. In logicians' jargon, the general proposition is the major premise of a syllogism (S39), which would read:

> All metals are elements.
> Iron is a metal.
> Therefore iron is an element.

Clearly, the argument excludes alloys such as pewter and brass, and refers only to pure or elemental metals. It can be said then, that the major premise is a tautology, so that it is true by definition, and we must accept the statement that iron is a metal, and therefore an element. But the major premise in some arguments can be derived only by induction (I8). For instance, the truth of the statement, 'This silk is the produce of XYZ Co. and is therefore pure' depends on the truth of an unstated major premise that all silk from XYZ is pure, which could not be true by definition. It could be proved, if at all, only by induction. As it stands, the inference that 'this silk is pure' is a *non sequitur* (N8). See statistical inference (S28).

D5   Deferred annuities

An annuity (A13) which comes into operation at some date in the future. The capital sum required now to purchase a deferred annuity of £1000 a year is less than the sum required to purchase an immediate annuity of £1000 a year. This follows from the fact that £ due at some date in the future is worth less than £ due today, the extent of the shortfall depending partly on the rate of interest and partly on the deferment period. At a rate of interest of 10%, the price of an immediate limited annuity of £1000 a year for 10 years is £1000 at £6.145 per £ = £6145. The figure of £6.145 per £ will be found in any published tables of the present value of £ a year (P23). The figure of £6145 is the present value of £1000 a year for 10 years beginning this year, with interest at 10%. If the annuity is to begin in five years' time, ie it is deferred for five years, then the price today would be the present value of £6145 due in five years:

Present value of £6145 due at the end of 5th year, with interest at 10%

£6145 at £0.6209 per £ = £3815

Where the purchase price consists of instalments paid during the period of deferment, the annual contribution is equal to the annual sinking fund (S18) required to provide the purchase price at the beginning of the first year of the annuity.

Conversely, if it is required to know the annuity deferred for five years, and limited to 10 years, which £3815 would purchase, the amount of £3815 (A8) at the end of five years is first calculated. According to published tables, the amount of £ at 10% in five years is £1.611, so that the amount of £3815 would be £6145, which would yield an annuity of £1000 a year for 10 years with interest at 10%

£6145 at 0.1627 per £ = £1000

The figure of £0.1627 is the reciprocal of the present value of £ a year. See A13.

Deferred perpetual annuities respond to the same kind of arithmetic. A perpetuity is a matter of simple interest: the purchase price of a perpetuity of £1000 at 10% is £10 000. There is no capital repayment involved, the annuity consisting entirely of interest. The present value of £10 000 due in five years with interest at 10% is

£10 000 at £0.6209 per £ = £6209

Applications. Investment appraisal, investment analysis, pension-fund calculations, efficiency auditing.

D6 Degrees of freedom

The number of sets of observations in excess of the number required to solve a problem by pure mathematics. It is usually written $n-m$, where n is the total number of sets of observations, and m the number required for solution by pure mathematics; this is equal to the number of unknown quantities to be calculated. The simple linear equation $Y = aX + b$, for which values of Y and X are available, has two unknowns: a and b. A solution by pure mathematics would call for the use of two values each of Y and X, ie of two sets of observations. Provided Y is not subject to random variations, and that the only factor influencing it is X, then such a solution would be perfectly satisfactory. But statistical methods assume the presence of random variations and of other factors, of minor significance though they may be. Then the parameters a and b can be calculated as averages only, subject to a margin of error; the greater the number of sets of observations in excess of the number of unknowns to be solved, the greater the reliance that can be placed on the averages. Hence the importance of the number of degrees of freedom.

Applications. Used chiefly internally in statistical analyses for correcting parameters for smallness of sample.

D7 Demand function

An equation used in the analysis of market demand. Marshall, in his *Principles of Economics,* first expounded a demand function. It is marginal (M3) in form and expresses the quantity in demand as a function of price:

$$Q = Kp^{-e} \qquad (7.1)$$

where Q is the quantity in demand, p the price, e the price elasticity of demand, and K is a constant. A form that applies to staple commodities whose supply is highly inelastic in the short run is

$$p = kQ^{-1/e} \qquad (7.2)$$

where $k = K^{1/e}$. But Marshall's diagrammatic form, in which Q is measured on the horizontal axis even where it is the dependent variable, has remained standard practice. Indeed it has important advantages, especially in break-even analysis and price-fixing theory, as Figures B14.1 and B14.2 clearly show.

Another form of demand function is the sales or gross-revenue function, in which the marginal form gives way to the aggregative form. In this, the sales proceeds, which reflect the effect of changes in price, take the place of price, and are measured on the vertical axis. With the same notation as above, and G for gross revenue, the function is

$$G = kQ^{1-1/e} \qquad (7.3)$$

Its graph rises from the origin to the right and is concave downwards, as curve S in Figure B14.2. Equation (7.3) is derived from (7.2) above.

For the purposes of regression analysis, the standard basic form of equation is (7.1) above, with other measurable relevant factors taken into account:

$$Q = Kp^{-e}S^{f}R^{g}D^{h} \qquad (7.4)$$

where S,R,D . . . represent other independent variables such as a price index of products in direct competition, consumer expenditure (or gross capital formation where Q applies to a capital good), the company's own advertising expenditure, competitors' advertising expenditure, and so on, and f, g, h . . . are the elasticities of demand with respect to these factors.

For regression purposes, the mathematical model is necessarily linear with respect to the parameters: it is the logarithmic form of (7.4):

$$\log Q = \log K - e \log p + f \log S \ldots \qquad (7.5)$$

General reasoning suggests that the appropriate type of model is of this type. Experience shows that it is also empirically valid. See M8 and !6.

Applications. Analysis of market demand; price fixing, investment appraisal, efficiency auditing.

D8 Demand schedule

A statistical schedule (S32) designed to show the effect of changes in price on the quantity of a brand of product sold in unit time, and on the sales proceeds. If the regression equation (R9) derived from an analysis of market demand gives an elasticity of demand, with respect to price, of 2, and a value of K of 100 000 a year, then

$$\log Q = 5 - 2 \log p$$

which is based on Equation (D7.1). For a range of prices from £10 to £15, the demand schedule would read as shown in Table D8.1. It should be kept in mind that the schedule holds all factors, other than price, constant at the average level of the period covered by the regression analysis. Any significant change in any factor would probably affect the value of the unattached constant, K, rather than the price elasticity of demand.

Table D8.1   Example of annual demand or sales schedule.

Price p (£)	p^2	Quantity $Q = \dfrac{100\ 000}{p^2}$	Sales proceeds $G = pQ$ (£)
10	100	1 000	10 000
11	121	826	9 086
12	144	694	8 328
13	169	592	7 696
14	196	510	7 140
15	225	444	6 660

Applications.   Net-revenue analysis, presentation purposes.

D9 Dependent variable

Of two or more factors involved in a causal relationship (C7), the one considered by general reasoning to be dependent on the other or others: the independent' variables (I6). As shown under C7, the nature of a causal relationship cannot be proved by statistical correlation (C25). In the field of management and economics generally, it is found that intercorrelation abounds, ie that B depends upon A and A depends upon B. For the purposes of regression analysis and multiple correlation, one of the factors has to be chosen to serve as the dependent variable.

D10   Depreciation

The reduction in the value of wasting fixed assets (F5) due to wear and tear in use and the passage of time; also used of the amount set aside periodically to make good the wastage, called *amortization* when the original outlay is to be recovered over the book-life of the asset—as in the case of leasehold land and property—and *renewals* or *renewal provision* when the company intends to renew the asset when it reaches the end of its life.

There are several methods of writing down the capital account to allow for the depreciation of fixed assets, the principal ones being:

(1)   The straight-line method, under which the net-replacement cost (R11) is divided by the number of years of book life, and the resulting figure written off the value of the asset at the end of each year.

(2)   The reducing-balance method, which is favoured by the Inland Revenue for taxation purposes. Under this, a percentage of the gross replacement cost (R11) is written off at the end of the first year, and the same percentage written off the balance remaining in each succeeding year. By the end of the book life of the asset, the balance remaining should approximate to the residual value of the asset.

(3)   The sinking-fund method (S18), which is nearly always a management-accountancy fiction, and not to be recommended for books of account. The annual sinking fund is rather less than the annual straight-line amount, since it assumes that the amount set aside annually accumulates at compound interest. Apart from the practice of prudent sole proprietors, the annual amounts set aside are generally used in the business, and where they are invested outside, the interest or dividend yield is accounted for separately in the financial books of account.

It is worth mentioning that depreciation is one of the most controversial subjects in the field of financial accountancy.

D11   Derived statistics

A generic term for statistical averages (A16), coefficients, other parameters (P3) and ratios. See basic statistics (B2).

D12   Design of statistical experiment
A statement of the purpose of, and proposed approach to an experiment or investigation involving statistical analysis. The approach covers the statistical methods to be used and a realistic form of mathematical model, which takes account of the relevant known and ascertainable basic and derived statistics.

D13   Deviation
The extent to which a figure of a series deviates from the arithmetic mean. See average deviation (A17), standard deviation (S24) and dispersion (D19).

D14   Diagram
A term that now appears to be used generically of all types of charts and graphs. A cake diagram (C1) is the same as a pie chart (P15), which is used entirely for presentation purposes; a scatter diagram (S9), sometimes called a dot diagram, is primarily an analytical tool of statisticians, though admittedly it is sometimes, but rarely, used for presentation purposes. Perhaps some attempt should be made to standardize the nomenclature, eg

Chart.   A visual means of presentation.
Diagram.   An analytical tool.
Graph.   The line or curve in a diagram depicting the relationship between two factors, such as the line of best fit (B5). Also see G5 and 6.

As graph paper was probably originally designed to assist mathematicians to draw graphs, the term may pass muster. But if the above definitions are accepted, the phrase *diagrammatic presentation* (D15) is self-contradictory; even so, it is so firmly established as an idiom — a piece of statisticians' jargon — that it must remain as an exception to the rule.

D15   Diagrammatic presentation
Presentation of statistics, by chart on a wall for general information, or in a book or article to illustrate an argument. The various kinds of chart are defined under their headings: B1, C1, P14 and 15, T6.

D16 Discounting

The reverse of compounding (C16). The appropriate formulae are those of the present value of £ and the present value of £ a year (P23).

Applications. Investment appraisal, cash-flow discounting, pension-fund calculations.

D17 Discrete variation

A step-by-step variation such as by whole indivisible numbers, as distinct from continuous variation (C21).

D18 Diseconomies of scale

An academic economic theory which argues that there is, in any industry, an optimum size of firm. Below the optimum size, economies of scale (E2) result from growth until the optimum is reached, when further growth results in diseconomies of scale. In an article on 'Making sense of growth', in the *Financial Times* of 19 September 1967, George Cyriax presented a table showing the size of firm in terms of sales turnover, and the ratio of profit to total assets, for a sample of firms in the UK. Firms with a turnover of up to £1 million a year had a profits ratio of 13.17%; those with one of £1-2 million, a ratio of 14.35%; £2-4 million, a ratio of 11.06%, a figure which remained fairly constant for all size groups over £4 million, the highest being 11.94% for £8-16 million. To what extent chance may have played a part in the variation of the profits ratio is not possible to determine from the data given, but the table presents a *prima facie* case for supposing that the optimum size of firm in the UK in 1965, to which the table relates, lay in the range £1-2 million turnover, diseconomies of scale occurring mainly during growth from £2-4 million.

D19 Dispersion

The extent to which the figures of a series vary. There are several measures of dispersion, the more useful ones being defined, or referred to below.

The *range,* the difference between the highest figure in the series and the lowest. It is open to the criticism that either one or the other, or both, of the two figures may be way out in the blue. To avoid such a possibility, the range may be measured between, say, the fifth largest and the fifth smallest. However, this is not satisfactory if the figure is to be compared with the range for a series with a different number of observations. For this reason, where range is considered to be an important indicator, the range between, say, the 90th *percentile* (P11) and the 10th percentile may be used. If the series consists of 140 observations, the difference between the figures for the 126th observation in order of magnitude and the 14th observation is the range; it would bear comparison with the range between the 144th and 16th observations of a series of 160 observations. See decile (D1).

A more popular range as a measure of dispersion seems to be the *interquartile range,* ie the difference between the upper quartile and the lower quartile (Q5), which is sometimes divided by 2 to give what is called the *quartile deviation.*

Two other measures of dispersion are the *average deviation* (A17) and the *standard deviation* (S24).

Applications.   Of wide use in statistical research and analysis; this is especially so of the standard deviation.

D20   Dividend %
The rate of dividend from shares in a company per £100 of nominal value. If the nominal price of a share is 50p and the annual declared dividend is 25%, each £100 of nominal value held would yield £25, subject to income tax, and the dividend amount per share would be 12½p. Declared dividends are net of corporation tax but gross of income tax, which is deducted at source. See D21.

D21   Dividend yield
The rate of dividend from shares in a company per £100 of market value. The formula is

$$\text{Dividend yield} = \frac{DN}{M}$$

where N is the nominal price, M the market price, and D the dividend %. For the numerical example in D20, and for a market price of £1 a share:

$$\text{Dividend yield} = \frac{25 \times 50}{100} = £12.50 \text{ per £100 of market value}$$

which is usually stated as 12½%.

E1   Earnings yield
The rate of company earnings per £100 of market value of its shares. Earnings, which play an important part in stock-market statistics (see P9, T7), may be defined as the total profit covering distributed and undistributed earnings.

E2   Economies of scale
See diseconomies of scale (D18).

E3   Efficiency ratios
Same as business ratios (B17).

E4   Eighty-twenty law
An empirically derived rule which suggests that 20% of the headings in a stores classification account for 80% of the value of stores, that 20% of customers account for 80% of the value of sales, that indeed, 20% of anything accounts for 80% of value, be it cost or revenue or assets.

E5   Employment function
An equation or mathematical model expressing the average labour force employed as a function of the scale of production, past experience of the product or products being produced, and the average size or weight of the product. In effect it consists of a form of the production function (P31), which expresses output in terms of quantity as a function of the average labour force, past experience and the average size of the product.

There is reason for supposing that the production function is of the logarithmic type. If it is written

$$Q = KF^a W^b E^c \qquad (5.1)$$

where Q is the output in unit time, K is a constant, F the average labour force, W the average size of product, E the cumulative output, and a, b and c are parameters which the analyst is seeking to evaluate, then the employment function reads:

$$F = K_2 Q^{1/a} W^{-b/a} E^{-c/a} \qquad (5.2)$$

where $K_2 = K^{-1/a}$. Although the analyst may prefer to use the logarithmic form of (5.1) for evaluating a, b, c and K by regression analysis, (5.2) provides the basis of the so-called percentage laws (P10) of scale and learning.

The formulae for the two laws are as follows.

Scale of Production Y%

$$Y = 50 \times 2^{1/a} \qquad (5.3)$$

Learning, X%

$$X = 100 \times 2^{-c/a} \qquad (5.4)$$

These percentage laws provide a simple means of presenting to non-mathematicians the significance of a regression equation based on (5.1) above. The former may be interpreted as meaning that every time the scale of production is doubled, the labour required per unit of output falls to $Y\%$ and the latter, that every time the cumulative output is doubled, the labour required per unit of output falls to $X\%$, other things being equal.

Applications. Forecasting, personnel planning, investment appraisal, efficiency auditing, annual costing.

E6   Economic ordering quantity (EOQ)

The optimum size of order for goods for sale or further processing, ie the size of order that minimizes the annual variable costs of carrying stock and of ordering. Cost of carrying stock covers:

1   Risk of obsolescence and deterioration.
2   Insurance of stocks, warehouses and plant.
3   Interest on value of stock.
4   Interest, depreciation and repairs to buildings and plant.
5   Handling labour.

Cost of ordering covers:

1   Internal cost of ordering.
2   Handling cost of reception, etc.
3   Vendor set-up cost (the differential discount for bulk deliveries).
4   Vendor set-up charge (the fixed charge per delivery).

In practice, some of these costs, e.g. the internal cost of ordering, may turn out to be non-variable, in which case they should be ignored.

Table E6.1   Statement of annual costs for different sizes of order or delivery.

Size of order (gal)	Price paid per gal (p)	Total per year (£)	Total set-up charges (£)	Stock-carrying costs Insurance (£)	Interest (£)	Total annual cost (£)
100	20	6 000	600.0	2.2	21.0	6 623.2
800	20	6 000	75.0	3.6	28.0	6 106.6
1 000	19	5 700	60.0	3.8	28.5	5 792.3
4 000	19	5 700	15.0	9.5	57.0	5 781.5
5 000	18	5 400	12.0	10.8	63.0	5 485.8
6 000	18	5 400	10.0	12.6	72.0	5 494.6
7 000	18	5 400	8.6	14.4	81.0	5 504.0
10 000	18	5 400	6.0	19.8	108.0	5 533.8

Attempts have been made to derive a mathematical formula expressing EOQ as a function of the stock carrying cost, the ordering cost, the annual usage and the vendor's quoted price. But, in practice, it seems the better approach is to build up a comparative statement of annual variable costs for different sizes of order. Table E6.1 contains a numerical example for a liquid chemical; the factory's annual consumption is 30 000gal, the quoted price 20p per gal, with a discount of 1p per gal for deliveries of 1000-5000gal, and 2p for deliveries of 5000gal and over. The vendor set-up charge is £2 a delivery. There is no risk of obsolescence or deterioration. The factory maintains a buffer stock of 1000gal, the cost of holding which is non-variable. The annual variable cost relevant to the EOQ is that part of the total annual factory cost that varies with the size of the order. Table E6.1 shows that the EOQ for the example is 5000gal, which, as might be expected, is the smallest size of order commanding the maximum discount. (The above numerical example is taken from E. J. Broster, *Planning Profit Strategies,* Longman, 1971, Table 17.1, where the text discusses the example in some depth.)

   Applications.   Stock control, investment appraisal.

E7   Error

Statisticians use the term *error* in the sense of difference, such as the actual figure minus the estimated figure, rather than in the sense of mistake, though they do take account of mistakes in measurement where they are relevant. See rounding error (R18) and standard error (S25).

E8   Exchange reserves

In effect the government's bank balance on overseas account. Before the pound sterling was floated in June 1972, the exchange equalization account operated the reserves to assist it in keeping the external value of the pound within its agreed limits. Drawings on, and additions to the reserves for the purpose all form part of what is now known as *official financing* of the balance of payments. However, it is only part, and it is unwise to judge the country's liquidity position solely by the magnitude of the reserves. By far the bulk of the official financing in recent years has been carried out by borrowing from, and repaying (including lending) to the International Monetary Fund and the monetary authorities of other countries.

E9   Exponential smoothing

An indefensible piece of jargon variously defined, but with an exponential equation as basis, derived on the assumption that the time series under review tends to vary from period to period in geometric progression. There are two forms of the basic equation:

$$Y = bR^n \tag{9.1}$$
ie
$$\log Y = \log b + n \log R \tag{9.2}$$
and
$$Y = b\,10^{an} \tag{9.3}$$
ie
$$\log Y = \log b + an \tag{9.4}$$

where n is the number of periods. In (9.1) and (9.2), R is the common ratio, so that $100(R - 1)$ is the percentage rate of increase. Since R is a constant in the equations, $\log R$ is also a constant and is clearly equal to a, which is also a constant, in (9.3) and (9.4). It follows that the common ratio is also equal to antilog a, and the percentage rate of increase is $100\{(\text{antilog } a) - 1\}$. It should be mentioned that since, where $n = 0$, $an = 0$, b can be said to be the norm for the period prior to the first period of the series. The values of log R or a, and b can be solved by simple-regression analysis (R7) based on either (9.2) or (9.4). Where the method of least squares (L7) is used, the appropriate formula is that of the regression of Y on n. See G2 and 6. An essential feature of exponential smoothing for forecasting is that later periods are given greater weight than earlier periods. See T16.

Applications. Forecasting, planning.

E10 Export List
A classification, with code numbers, of the commodities entering into the visible export trade of the UK. It also contains a list of countries of destination, and is revised annually, the new revised List becoming available for the use of exporters from HMSO by 1 January of each year.

E11 Export statistics
See overseas trade (O8)

E12 Extrapolation
The extension of a graph beyond the upper or lower limit of the observed range of the basic statistics, or the calculation from a regression equation (R9) of an estimated value or values of the dependent variable corresponding to a value or values of one or more of the independent variables outside the observed range. It is a process which theoretical statisticians frown upon, and indeed, it needs using with care. The further the graph is extended beyond the observed range the more dangerous the process becomes; the same applies to extrapolated values calculated from a regression equation.

Applications. Forecasting, planning.

F1   Family-budget inquiry

A sampling exercise (S3) designed to determine the amounts that the average family spends in a period on the various items of family expenditure including holidays, travel, eating out, rent, rates and the amount saved in the period. The Department of Employment and Productivity conducts a family-budget inquiry from time to time to keep the pattern of weights for the retail-price index up to date. See C30.

F2   Fiducial limits

Sometimes called *fiduciary limits* or *confidence limits;* the limits within which a derived figure such as an average or a standard deviation (S24) is true of the universe (U3). The fiducial probability that the true average of the universe falls within the fiducial limits set by the standard deviation of the statistical series applied to the average of the series, is 68%. One-and-a-half times the standard deviation has a fiducial probability of 87%, and twice the standard deviation has one of 95%. If the average of a series is 156, and the standard deviation is 20, then the fiducial probability is 68% that the true mean of the universe falls within the fiducial limits of 156 ± 20, ie between 136 and 176. See B4 and F9.

Applications.   Sampling, forecasting, quality control.

F3   Finite differences, method of

Sometimes called the *calculus of finite differences.* A method of simple-regression analysis in its own right, but more usefully applied in double harness with the methods of cross-classification (C36) and group averages (G9). As a method in its own right, it is useful for acquiring the 'feel' of the basic data. First, the data are arranged in order of magnitude of the independent variable, then either the successive differences are extracted, or the deviations from the mean. A selection of 10 sets of observations limited to miles and man hours from the example of Table C36.1 can be used for demonstrating the method in its application to linear types of model. Table F3.1 gives

Table F3.1   Finite successive differences applied to the example.

M	H	M^1	H^1	H^1/M^1
50	40	–	–	–
80	49	30	9	0.300
90	54	10	5	0.500
106	60	16	6	0.375
110	65	4	5	1.250
118	67	8	2	1.000
120	69	2	2	1.000
121	68	1	−1	−1.000
124	72	3	4	1.333
128	71	4	−1	−0.250
		78	31	0.398

the selection and the successive differences approach to a solution. The figures in the righthand column are estimates of the value of the regression coefficient of M in the equation $H = aM + K$, which ignores the number of calls made as an independent variable. It will be seen that the estimates range from −1.000 to +1.333 with an apparent average of 0.398, to which little significance should be attached since it is in effect based entirely on the first and last sets of observations in the table, ie $128 − 50 = 78$, and $71 − 40 = 31$. If the second and penultimate sets were used, we would have: $124 − 80 = 44$ and $72 − 49 = 23$; $23/44 = 0.502$, which happens to lie more closely to the figure 0.4681 in Table C36.1 than is the figure given above. If we chose to base the solution value of a on the average of the first half of the series and of the second half, we would be harnessing finite differences to the method of group averages (G9).

The mean-deviations approach is similar to the successive-differences approach, the difference being that the values

61

of M^1 and H^1 are the mean deviations, and there are 10 of them in this example instead of 9. The values of N^1 and H^1 furthest from the middle of the series give the more significant solution estimates of a.

Finite differences as a regression method in its own right has its most valuable application in the solution of problems in two variables where the basic model appears to be an equation of the second or higher degree:

$$Y = K + a_1 X + a_2 X^2 + a_3 X^3 \ldots + a_n X^n$$

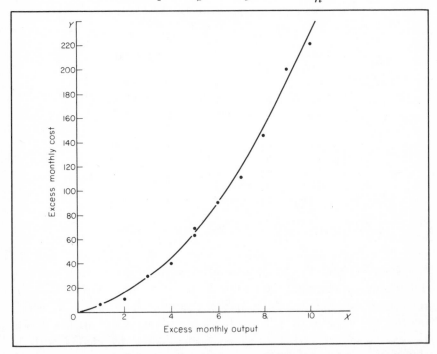

Figure F3.1   A typical total cost curve for output in excess of the normal plant capacity

The observed values of Y are plotted on rectilinear graph paper against the corresponding values of X, and to the scatter of dots thus obtained, a smooth curve is fitted freehand, ie by personal judgement. About eight sets of readings are then taken from the curve. It is these readings, not the basic data themselves, that are processed by finite differences. Suppose the rate of output of a firm's product is exceeding the normal capacity of the plant (P17), and the management accountant requires a mathematical model to facilitate cost estimation and forecasting. He has monthly statistics of excess output and excess variable cost for a period of 11 months as follows:

Excess output X, 1 3 5 4 2   7 5 6   9   8 10
Excess cost Y, 7 30 70 40 10 110 63 90 200 145 220

He plots these − as in Figure F3.1 − and fits, by personal judgement, a smooth curve to the scatter of dots. He thinks an equation of the second or third degree might satisfy the curve, and decides to proceed by finite differences to test his conclusion and, if this is valid, to derive the required parameters. Eight readings from the smooth curve are given in rows (a) and (b) of Table F3.2. The procedure used for completing the first stage is shown in the table.

Table F3.2   Analysis of curve in Figure F3.1 by finite differences.

Column line	(1)	(2)	(3)	(4)	(5)	(6)	(7)	(8)	(9)
Readings from graph									
(a)	X	1	2	3	4	5	6	8	10
(b)	Y	6	15	28	45	66	91	153	231
First differences									
(c)	X'	−	1	1	1	1	1	2	2
(d)	Y'	−	9	13	17	21	25	62	78
(e)	Y'/X'	−	9	13	17	21	25	31	39
Second differences									
(f)	X''	−	−	2	2	2	2	3	4
(g)	Y''	−	−	4	4	4	4	6	8
(h)	Y''/X''	−	−	2	2	2	2	2	2

Notes Lines (c) and (d): the successive differences in lines (a) and (b), respectively.
Line (f): from line (a): $3-1, 4-2, \ldots, 10-6$.
Line (g): the successive differences in line (e).

Since row (h) of Table 3.2 is trendless, the appropriate equation is a quadratic, and since the calculated values of Y''/X'' are all 2, the coefficient of X^2 is 2. We now have

$$Y = K + a_1 X + 2X^2$$

Since the calculated values of Y''/X'' are all the same, indicating a perfect fit, the values of K and a_1 can be derived from any pair of sets of readings after eliminating $2X^2$ from Y. The complete regression equation is

$$Y = 1 + 3X + 2X^2$$

Needless to say, perfect fits in practice are few and far between. Readings from graphs are subject to a margin of error, but if that were all, one or two slight adjustments of the readings could very well prove worthwhile. If the Y''/X'' series still reveals a trend, the process needs to be extended to third differences, when, if the Y'''/X''' series turns out to be trendless, the appropriate equation would be one of the third degree. An increase in trend would indicate that the appropriate equation is not of the type, and the analysis would have to be abandoned.

In practice, eight readings from the smooth curve are usually enough, but where there is a chance that the equation is of the fourth degree or higher, it would be wise to take 10 or more readings for the analysis.

Applications. Cost analysis, model building, regression analysis.

F4 Fisher's ideal index number
The geometric mean of Laspeyres's and Paasche's index numbers (L4 and P1). Laspeyres and Paasche are biased, if at all in opposite directions; where the index is one of prices, the former usually upwards and the latter downwards. Professor Irving Fisher, the American economist and author of the standard work on index numbers, *The Making of Index Numbers,* suggested that a cross of Laspeyres and Paasche would therefore give the ideal index number. The subject is more fully discussed under index numbers (I7).

F5 Fixed assets
Those capital assets that are durable and not easily lost. A connoted definition is buildings, machinery, plant, vehicles, ships, boats, aircraft and goodwill. Loose tools, including small portable power tools, materials, work in progress and finished goods, do not count as fixed assets. There are two dichotomies: tangible and intangible, and wasting and non-wasting. Intangibles consist mainly of goodwill, patents and copyright; non-wasting fixed assets, of freehold land and earthworks.

F6 Fixed-base index number
A series of index numbers with a common base (= 100). The base usually covers one of the periods, times or places within the series, not necessarily the first in a time series. It serves as a common

standard of comparison. The weighting system does not necessarily follow the base. The pattern of weights applies to the base period or place when Laspeyres's index is used, and to the given period or place when Paasche's index is used. See L4 and P1. A chain-base index (C11), applicable only to time series, uses the weights either of the given period or of the preceding period, the base always being the preceding period.

F7   Fixed capital formation (FCF)

The addition in unit time to the stock of the fixed capital of an undertaking, a region, a country or the whole world, in terms of value. For an undertaking, it may include purchases in the period of freehold land and, sometimes, acquisitions of leasehold land. For a country, it would exclude land except where it had been won from the sea or acquired from another country by purchase or conquest.

Gross FCF is the total acquisition of fixed capital in the period, regardless of the wastage or depreciation of existing fixed capital.

Net FCF is the total less depreciation, or the total less that part of the fixed capital assets acquired in the period to replace fixed capital assets desplaced. This net figure is sometimes called betterment (B7).

In the national accounts, the gross FCF, described as the gross domestic fixed capital formation, is defined by connotation as 'expenditure on fixed assets (buildings, vehicles, plant and machinery, etc.) either for replacing or adding to the stock of existing fixed assets. Expenditure on maintenance and repairs is excluded'.

Sources.   UK figures: *National Income and Expenditure,* annual blue book, HMSO, for annual statistics; *Economic Trends,* monthly, HMSO, for quarterly statistics. The blue book gives both gross and net FCF and a number of analyses, eg by sector, industry and type of asset.

F8   Fixed costs

Annual costs that remain unaffected by some entity under consideration, such as the rate of output, the capacity of the plant, the labour force, the management team, the stocks of materials and so on. When used unqualified, it usually refers to output, ie the term means costs that are not influenced by changes in the rate of output. Sometimes called *non-variable costs.* See attribution costing (A15), marginal cost (M4) and variable costing (V5).

Applications. Break-even analysis, profit planning, product costing, project costing.

F9   Frequency distribution

A table showing the number of items that fall within each of a number of equal ranges or class intervals of a series of figures distributed by order of magnitude. The number of items in a range is the frequency. An example of a frequency distribution is given in Table D3.1: column 2 shows the ranges by reference to the middle figure of each, and column 3, the frequency. It is only with continuous series (C21) that the range is strictly necessary. Where the series is discrete (D17), as it is with that of the example of the meat pies in Table D3.1, the basic series may progress unit by unit, ie the range, so to speak, may consist of one unit. Since the daily sales of meat pies are distributed by the number of days out of 100, the anticipated number of days may be regarded as the percentage probability that so many pies will be sold.

A frequency curve is a smooth curve fitted to the points of a continuous series, with ranges measured on the horizontal scale, and the frequency on the vertical scale. Where the series is discrete, the points are connected by straight lines, and the diagram is then called a frequency polygon. An example of a frequency polygon is given in Figure D3.1.

Frequency curves (and polygons) are of three broad kinds: (1) the symmetrical or bell-shaped curve; (2) the righthand or negatively skewed curve; (c) the lefthand or positively skewed curve. There is another trichotomy: (1) normal; (2) pointed; (3) flat. The normal frequency curve has a mathematical foundation, proved valid empirically by numerous experiments. It is fundamental in the statistical theory of probability. If a length of road is measured 100 times by different people using different types of equipment, the mode, median and arithmetic mean of the measurements would all be equal to each other, and they would tend to be the correct measurement. The number of measurements falling short of the mode would be the same as the number exceeding it, and the number of the more extreme measurements — those very low and those very high — would be smaller than the number of more nearly accurate measurements in ranges of the same magnitude. The frequency curve of the measurements would conform to the normal frequency curve, which is variously called the *normal curve of error* and the *normal probability curve* (N9).

The normal frequency curve has some important properties, which have made it possible to calculate the fiducial limits (F2) of the standard deviation (S24), and other probabilities. There is a 68% probability that any observation will fall within one standard deviation of the average, and therefore a 32% probability that it will fall outside it. This means that if the frequency distribution of the 100 road measurements conformed to the normal curve of error, which it would undoubtedly do, 68 measurements would fall within one standard deviation of the average. The fiducial probabilities relating to fractions and multiples of the standard deviation are given under normal curve of error (N9), where the subject is examined in some depth.

It is remarkable how, in practice, symmetrical frequency distributions conform to the normal curve. Even short series are often found to conform near enough. Of many series of 15 observations, for instance, 10, or 67%, fall within one standard deviation of the average, with 14, or all 15, falling within two standard deviations of the average.

Applications.   Making (or growing) decision trees, and estimating standard errors and errors of estimate.

F10   *F* test

A test of the significance of the difference between sample variances (V7); sometimes called the *variance ratio test. F* is equal to the larger estimate of the variance of the universe divided by the smaller estimate. See universe (U3), analysis of variance (A10) and variance (V7). The variances are those of an assumed universe (from which two samples are drawn) postulated by a null hypothesis (N11), the two variances being equal to the squares of the two sample standard deviations with Bessel's correction (B4). It will be appreciated that the greater the value of *F,* the smaller the amount of faith one can place in the null hypothesis. To test the significance of *F,* a set of statistical tables, known as Snedecor's tables, have to be consulted. They give the percentage probabilities that the value of *F* arises by chance, and therefore indicate whether the null hypothesis is valid or not.

F11   Function

In statistics used in the mathematical sense, that is, to quote the Oxford Dictionary, 'variable quantity in relation to other(s) in terms of which it may be expressed or on which its value depends'. Mathematicians and mathematical statisticians use a form of short-hand, eg

$$Y = f(X)$$

which simply means that Y is a function of X. In practical statistics, an equation like this serves merely as a preliminery model: it saves writing out a verbal model. Before it can be applied in practice, it has to be converted to an equation which indicates the type of relation-ship that exists theoretically, or empirically, between the variable quantity and the other(s), eg

$$Y = aX + b$$
$$Y = bX^a$$
$$Y = aX^2 + bX + C$$
$$Y = ba^x$$

See model building (M16), also C28, D7, E5 for types of function of use in management statistics.

F12   Future value

Terminal valuation (T3).

g

G1   Gearing

Often called *capital gearing;* the ratio of fixed-interest capital (preference shares, debentures and other loan debt) to the total capital employed in a company; usually expressed as a percentage. Its object is to indicate the capital structure of a company, and although it does this, it is sometimes used as one of the factors in evaluating the ordinary shares in a company. Other things being equal, where the percentage is high, the equity earnings and dividends are more sensitive to changes in economic conditions and similar factors than where it is low. However, the sensitiveness arises not from the gearing ratio but from the capital-cost ratio. Equity earnings and dividends would be more sensitive to a 10% debenture issue of £100 000, than to a 4% debenture issue of the same amount: the former costs the company £10 000 a year, the latter, £4000 a year. For the purpose, one might use the ratio of the annual cost of fixed-interest capital to the current total cost of capital. Another criterion is the ratio of the current market value of the company's fixed-interest capital to that of the total issued capital and loan capital. The possible effect on the gearing ratio may be judged from a particular case. According to the *Stock Exchange Investment List,* October 1972, the firm Crosfields and Calthrop classified in the FT share information service under 'Food and groceries', had the quoted capital issues shown in Table G1.1.

For the cost-of-capital criterion, it is necessary to take account, not only of the dividend and interest cost, but of any special conditions attached to the issue of the fixed-interest stocks. The definition of the word *convertible* depends on the terms and

Table G1.1   Gearing ratio, nominal and market-value criteria.

Stock or share	Nominal value £	Mid-market price per £100 on 18 September	Market valuation (£)
A Ordinary at 25p	1 397 100	220.0	3 073 620
B 4% Cumulative preference	150 000	37.5	56 250
C 9½% Convertible loan stock 1986-91	1 250 000	105.0	1 312 500
Total capital	2 797 100		4 442 350
Total fixed-interest capital	1 400 000		1 368 750
Ratio of fixed-interest to total, %	50.05		30.81

conditions of issue. For the purposes of the argument, the convertible loan stock can be assumed to be redeemable at par in 1991, say 20 years after issue. This would mean that, in addition to the interest charge of 9½%, the company would need to set aside each year a sum which, accumulating at compound interest, would provide £1 250 000 in the year 1991. This would be the annual sinking fund (S18), which, at 9½% for 20 years, is £0.01847 per £ giving a total annual sum of £23 095.

Table G1.2   The cost-of-capital criterion.

Stock	Nominal value (£)	Rate of dividend or interest, % (£)	Annual cost of capital (£)
A	1 397 100	8	111 768
B	150 000	4	6 000
C	1 250 000	9½	118 750
Redemption of C: annual sinking fund			23 095
Total annual cost of capital			259 613
Total cost of fixed-interest capital			147 845
Ratio of fixed-interest cost to total, %			56.95

Table G1.2 gives the necessary calculations for the cost-of-capital criterion. It might be argued that it would be more correct to apply the gross redemption yield to the amount of the loan stock instead of charging interest plus sinking fund separately.

It will be seen from the two tables that the three criteria may give entirely different ratios. The rate of dividend is crucial in such comparative analyses, and since it varies enormously from company to company, affects the comparison appreciably. It also varies from time to time. Crosfields and Calthrop have indicated that the next full year's dividend will be 10%. This would give a total annual cost of ordinary share capital of £139 710, a total annual cost of all capital of £287 555, and a gearing ratio of 51.42%; which is significantly smaller than the 56.95% based on a dividend of 8%, shown in Table G1.2.

Applications. Investment analysis, investment planning.

G2 Geometric mean (GM)

The GM of a series of figures is the nth root of their product, where n is the number of figures in the series. In practice, where n exceeds 2, it is usual to extract the nth root by using logarithms. A practical formula is therefore

$$GM = antilog \frac{\Sigma(\log X)}{n} \qquad (2.1)$$

It is of little use either as a measure of location or central tendency: it has a downward bias compared with the arithmetic mean. The AM of 2 and 8 is 5, whereas the GM is 4; where one or more of the figures in a series is zero, the GM is also zero. The AM of 2, 7 and 0 is 3; the GM is 0. Logarithms do not help, since the logarithm of zero is minus infinity, and the sum of a series of numbers, one of which is minus infinity, is minus infinity.

However, the GM is not without its uses. For instance, populations tend to grow in geometric progression: where the population of a town is recorded as 40 000 in 1961 and as 90 000 in 1971, the likely population in 1966 (the middle year of the period) is the GM of 40 000 and 90 000, ie 60 000. To estimate the population in any other year, it would be necessary first to ascertain the common ratio. It is shown in A8 that the amount, A, of £X invested at r per £ interest for n years is

$$A = X(1 + r)^n \qquad (2.2)$$

The term $1 + r$ is the common ratio. From this, we have

$$1 + r = \left(\frac{A}{X}\right)^{1/n} \tag{2.3}$$

For the population problem, we therefore have

$$1 + r = \left(\frac{90\ 000}{40\ 000}\right)^{1/10}$$

$$\log (1 + r) = \frac{1}{10} \log 2.5$$

$$= 0.03522$$

After the lapse of five years, with 40 000 counting as the first term of the geometric progression, we have an estimate of the population of the town in 1966, of

$$40\ 000 \text{ antilog } (0.03522 \times 5)$$

ie $\qquad 40\ 000 \times 1\ 500 = 60\ 000$

which agrees with the GM.

There are 11 terms, not 10, in the complete geometric progression. It follows that the correct formula for determining intermediate terms of a geometric progression is

$$A = X(1 + r)^{n-1} \tag{2.4}$$

ie $\qquad \log A = \log X + (n - 1) \log (1 + r) \tag{2.5}$

where A is the value of the nth term, X the value of the first term, and $1 + r$, the common ratio, $100r$ being the percentage rate of increase. In the example, we have $\log (1 + r) = 0.03522$, so that $1 + r = 1.084$, and $100r = 8.4\%$ per annum.

Applications. Regression analysis of time series involving logarithmic models; forecasting, planning.

G3   Goodness of fit
The goodness of fit of a regression equation is measured by the standard error of estimate. See standard error (S25).

G4   Grand average
The average of a universe (U3) as distinct from that of a sample (S3) of the universe.

G5 Graphing
Plotting a scatter or dot diagram and fitting a graph to the scatter of dots. See line of best fit (B5).

G6 Graphs, types of
In determining empirically the kind of relationship that exists between two factors, model building (M7) and similar exercises, it is often useful to have a diagram of the various simple types of curvilinear graphs, apart from quadratics, that exist. There are two kinds of mathematical model of use for this purpose in regression analysis. They are

$$Y = bX^a \tag{6.1}$$

of which the logarithmic form is

$$\log Y = \log b + a \log X$$

where Y and X represent the two factors and

$$Y = b \times 10^{aX} \tag{6.2}$$

of which, since $\log 10 = 1$, the logarithmic form is

$$\log Y = \log b + aX \tag{6.3}$$

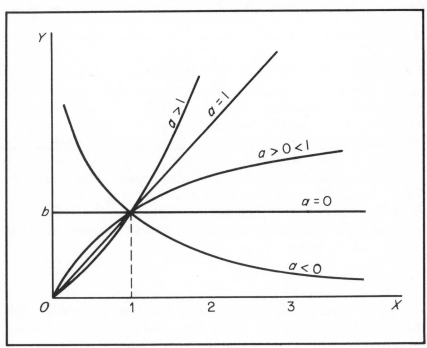

Figure G6.1   General shapes of graph of $Y = bX^a$

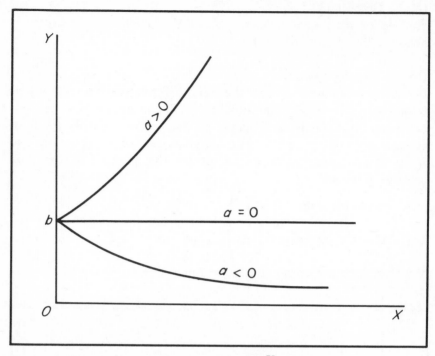

Figure G6.2   General shapes of $Y = b10^{aX}$

Figures G6.1 and G6.2 show that the general shapes that the graphs of these two equations may take varies with the value of a in both. In the former, there are two straight-line graphs, one for $a = 0$ and the other for $a = 1$, and in the latter, there is one straight-line graph, that for $a = 0$. All three are inserted to complete the picture. For where $a = 0$, $y = b$ in both cases, and the correlation between Y and X is zero. See C9. In the former case, for $a = 1$, the relationship is linear, ie $Y = bX$, and should be treated accordingly.

Applications.   Analysis of market demand; deriving the production and employment functions, regression analysis.

G7   Gross redemption yield
Officially defined as that rate of interest which discounts the future interest yield and the sum payable on redemption to a present value (P23) equal to the current market price of the stock. It 'comprises the gross flat yield together with an apportionment of the capital gain or loss on dated securities held to redemption'. *Financial*

Statistics, Notes and Definitions, HMSO, April 1969.

Anyone familiar with the internal rate-of-return technique of discounted cash flow will see that the gross redemption yield is much the same as the IRR (I14). It is almost as difficult to calculate, too. If a 3% stock maturing in 10 years stands at £67.1 per £100, its flat interest yield would be 4.47%:

$$\frac{3 \times 100}{67.1} = 4.47\%$$

and its gross redemption yield, 8%:

	£
Present value of £3 a year for 10 years at 8%	
£3 at £6.710 per £	20.8
Present value of £100 receivable in 10 years at 8%	
£100 at £0.463 per £	46.3
	67.1

The two rates per £ of £6.710 and £0.463 are taken from published compound-interest tables (*Inwood's* or *Parry's*).

If the current market price had been £100, which is called the parity price, then both the flat rate of interest and gross redemption yield would have been 3% the same as the coupon rate:

	£
Present value of £3 a year for 10 years at 3%	
£3 at £8.530 per £	29.59
Present value of £100 receivable in 10 years at 3%	
£100 at £0.7441 per £	74.41
	100.00

Where the price exceeds parity, the flat rate is less than the coupon rate, and the gross redemption yield is less than the flat rate. In these days of very high interest rates, the gross redemption yields, even of gilt-edged securities, rarely fall below 7%, and then only for securities with no more than a few months to go to maturity. In November 1972, longer-dated gilt-edged securities had gross redemption yields ranging up to more than 9½% and in April 1973, to more than 10%.

Applications. Investment analysis.

G8   Grossing up

A process used for estimating the total for a universe (U3) from a sample whose proportion of the universe is known precisely or approximately. The multiplier used for the purpose is known as the *grossing-up factor,* which, for a 5% sample, would be 20.

G9   Group averages

A method of simple-regression analysis (R1), in which the sets of observations are arranged in two or more groups of equal size in accordance with the magnitude of the independent variable (I6), averaged, and either plotted in a graph (G5), usually with the individual items, or analysed by finite differences (F3). See line of best fit (B5) and cross-classification (C36), under which the method of group averages in tandem with finite differences is demonstrated in Table C36.3. The term *group averages* is used in reference to simple regression, the term *cross-classification* – which is also a method of group averages – in multiple regression. Both are methods of regression analysis in their own right, as is shown in C36, but their chief value lies in empirical model building (M16) to provide a basic equation for use with least squares (L7).

Applications.   Regression analysis.

G10   Group internal-control matrix

A matrix or table designed to facilitate the control of the inter-divisional transactions of a company or the intercompany trans-actions of a group. Basically, it consists of an input-output matrix (I9) internal to the company or group. A hypothetical example is contained in Table G10.1. *A, B* and *C* are three manufacturing and trading divisions of a company. Each manufactures and assembles a

Table G10.1   An example of a group internal-control matrix.

Internal transactions, £'000

Purchases by	A	B	C	Total	External Sales	Total Sales
Sales by						
A	–	8	42	50	20	70
B	8	–	4	12	20	32
C	12	8	–	20	60	80
Total	20	16	46	82	100	182
External purchases	10	4	14			
Total purchases	30	20	60			

single product for external sale, and manufactures certain parts that are common to the products of all three divisions.

Suppose it is planned to double the external sales of A and to leave the external sales of the other two divisions unchanged. In consequence, the demand for common parts by A on B and C would also double, that on B from 8 to 16, and that on C from 12 to 24. The planned outputs of B and C would need to be increased accordingly; the same applies to A's external purchases.

In practice, however, the situation is rarely so simple: transactions often consist in part, at least, of general trafficking. If the three divisions produce and sell much the same kind of product, then, if A's external sales are doubled, its total sales would increase from 70 to 90, ie by 28.6%, and A's demand for supplies from B and C would tend to increase by the same percentage, those from B rising from 8 to 10.3, those from C, from 12 to 15.4. Then we would have the following changes in the total sales of B and C: B's by 2.3, from 32 to 34.3, ie by 7.20%; C's by 3.4, from 80 to 83.4, ie by 4.25%. In consequence, B's demand from A would tend to increase by 7.20%, from 8 to 8.6, that from C by 7.20%, from 8 to 8.6; C's demand from A would tend to increase by 4.25%, from 4 to 4.2. Thus, the total sales of each would increase further, and there would tend to be a further demand for supplies each from the other. This is an example of a chain reaction that theoretically could continue indefinitely.

The two above examples of the application of the group internal-control matrix demonstrate the two extreme situations. In practice it is likely that some intermediate position would be found.

Applications.   Forecasting, planning, internal control.

h

H1   Harmonic mean (HM)

An inverted form of the arithmetic mean applicable to the averaging of certain ratios, such as miles per hour and miles per gallon of fuel. Needs using with care, and in cases of doubt, the arithmetic method of extension with the arithmetic mean should be used. If a new car does 40mpg for its first 4000 miles, and 30mpg for its second 2000 miles, the average mpg over the 6000 miles can be calculated by using the HM, for which the formula is

$$HM = \frac{\Sigma n}{\Sigma(\frac{n}{x})}$$

where n is the weight allotted to each item; Σ the sign of summation, and x the values of the items to be averaged. The weight is represented by the numerator of the ratios being averaged, in this case, the mileage. Using units of 1000 miles for weighting purposes, we have

$$HM = \frac{4 + 2}{(4/40 + 2/30)} = \frac{6}{(20/120)} = \frac{720}{20} = 36mpg$$

The arithmetic method of extension can be demonstrated to check the figure of 36mpg.

Petrol consumption in first 4000 miles $= \dfrac{4000}{40} = 100$

Petrol consumption in next 2000 miles $= \dfrac{2000}{30} = 66.667$

Total consumption in 6000 miles = 166.667

Average mpg over 6000 miles is therefore

$$\frac{6\ 000}{166.67} = \frac{18\ 000}{500} = 36mpg$$

Here the numerator and denominator are both multiplied by 3 to rid the latter of the recurring decimal.

The weighted arithmetic mean of 40mpg and 30mpg is calculated as follows:

$$40 \times 4 = 160$$
$$30 \times 2 = 60$$

Total   6   220

$$AM = 220/6 = 36.667$$

which, as the above extension shows, is incorrect.

Note that many textbooks give the formula, not of the weighted, but of the unweighted HM, which is useful where the pattern of weights is not available. Strictly, the so-called unweighted average is an average in which the items are given equal weight. If unity, or 2 or any other figure, for that matter, is used as the common weight, the formula above may be employed for determining the unweighted average. See weighted average (W2).

H2   Histogram
The diagram of a frequency distribution (F9).

H3   Historigram
The graph of a time series (T8 and 9) with the series measured on the vertical scale, and time measured on the horizontal scale. It is common practice to connect adjacent dots by straight lines and not to attempt to fit a smoothed curve through the dots themselves, though a general curve can be fitted by regression analysis (R7) or by moving averages (M21). The distance between scale points on the time scale represents the lapse of time rather than periods, so that lines drawn from the calibration marks of the time scale should represent either periods or points of time for rate-of-flow figures and point-of-time figures (P18), respectively; the plotted dots thus fall on the lines and not between them.

Applications.   Planning, forecasting.

H4   Holiday relief
A cost-accounting term for the cost of annual leave. For holidays of one week a year, the addition to the cost of providing a relief is 1/51 of the cost, since the relief himself takes a week's holiday and is therefore available for only 51 weeks of the year. Similarly, for

holidays of a fortnight a year, the addition is 1/25, and for a calendar month, it is 1/11.

H5   Hybrid series

A statistical series consisting of a mixtue of time series figures (T8) and cross-section figures (C37), sometimes called more precisely a *temporal cross-section series*. It is adopted for analytical purposes where: (1) enough data of one kind or the other are not available; (2) cross-section data are limited in number of observations and have to be supplemented by time data; (3) in a regression analysis, cross-section data are preferred but are available for some variables only, and time series data have to be used for other variables. In the latter case, two separate analyses are carried out, and the two resulting regression equations merged into one.

H6   Hypothesis

A working theory forming the basis of an investigation. A written hypothesis consists of a statement of (1) the objective; (2) the several approaches that may be taken to find a solution; (3) the approach likely to be the most fruitful. (4) the available relevant information; (5) the additional information that would be required; (6) the limits set on the achievement of an acceptable solution by the information available or likely to be obtainable. The answers to these points may be described as the fundamentals of the hypothesis. The final stage depends on the objective and the alternative approaches to a solution. If the objective is to obtain an estimate of the price elasticity of demand for one of the company's products, for instance, there would be three possible approaches:

1   A market-research project (M7) in a selected test area.
2   Statistical analysis of the opinions of a selection of market executives and salesmen, of the effect on demand of given changes in the price of the product. See sales analysis (S1).
3   A regression analysis (R7). See demand function (D7).

The statistical data required for a regression analysis may not be available or obtainable, and a market-research project may be considered too expensive. This would leave an analysis of market men's opinions as the only feasible approach.

Experience shows that a carefully prepared hypothesis may ultimately save a great deal of time and expense. See also design of experiment (D12); for null hypothesis, see N11.

Applications. Research work of all kinds.

i

I 1 Imperfect competition

A degree of competition which affords the seller an appreciable amount of control over his selling prices. A measure of the degree of competition is the elasticity of demand, which is infinitely high where competition is perfect, and finite where competition is imperfect; the more imperfect, the lower the elasticity of demand. Imperfect competition arises from the monopolies existing in branded goods: the Bovril company, for instance, holds a complete monopoly of 'Bovril'; but the presence in the market of close substitutes like 'Oxo' and 'Marmite' tends to make the elasticity of demand somewhat higher than it otherwise would be. See branding (B13), perfect competition (P12) and rational price fixing (R7); also break-even analysis (B14).

I 2 Implied elasticity of demand

A null hypothesis (N11), to the effect that the current price of a brand of a firm's products is rational (R7) based on the postulation that the implied price elasticity of demand is the true one. The formula of the rational price (R7) or *optimum price,* as it is often called, is

$$p_m = a\frac{e}{e-1} \tag{2.1}$$

where p_m is the rational price, a the marginal cost (M4), and e the elasticity of demand. We therefore have

$$\frac{e}{e-1} = \frac{p_m}{a} \tag{2.2}$$

and

$$e = \frac{pm}{p_m{-}a} \qquad (2.3)$$

Let p be the actual price, whether rational or not, and e_i, the implied elasticity of demand, then

$$e_i = \frac{p}{p-a} \qquad (2.4)$$

It will be seen from (2.3) and (2.4) that, in a generally rational situation, the value of the elasticity of demand always exceeds unity, and that the higher the price in relation to the marginal cost, the lower the implied elasticity of demand. Equation (2.1) shows that the rational price varies *pro rata* with the marginal cost, so that it is possible to draw up a table for a range of elasticities of demand showing the rational price per money unit of marginal cost, eg

Elasticity of demand	Rational price
1.05	21.000
1.1	11.000
1.2	6.000
1.3	4.333
7.0	1.167
7.5	1.154

Experience suggests that the elasticity of demand for individual brands ranges from about 1.10 to about 4.00, though no known research into the problem has been carried out. A calculated implied elasticity of demand outside this range is likely to differ from the true elasticity of demand. A price corresponding to an implied elasticity of less than 1.1 is likely to be too low, and one corresponding to an implied elasticity of more than 4.00 is likely to be too high. However, experience with other brands of the company's products is the best guide to the reasonableness of an implied elasticity of demand.

Applications. Pricing policy, especially in respect of new product brands.

I 3 Import List
The old official, and still popular, title of what is now officially the *Statistical Classification for Imported Goods and for Re-exported*

Goods. The change from *Import List* was made in 1959, oddly enough, some years after the publication by HMSO of Sir Ernest Gower's *Plain Words.* Like the *Export List* (E10), the Import List is revised annually, the new edition becoming available by 1 January of each year. It contains a classification of goods entering into the import trade of the UK, and a list of countries of origin. It is officially described as complementary to the statistical key in Part 3 of *HM Customs & Excise Tariff,* which also contains information concerning the entry of imported goods, forms to be used, and other matters.

I   4   Incremental analysis
The process of determining an optimum value (O6) by building up a statistical schedule (S32). Examples of the process are given under decision tree (D3), EOQ (E6), net-revenue schedule (N6) and sales schedule (S2).

I   5   Incremental profit
1   The effect on net revenue (N6), of a change in one of the factors influencing it, eg sales turnover and selling price.
2   The effect on the net cash flow (C6) of a change in one of the factors influencing it, or of each of a number of courses of action compared with the other possible courses of action.

Net cash flow is defined for the purpose as the difference between the actual inflow of money and the actual outflow of money. Incremental profit, under its second definition, is the measure of the potential change in the company's liquidity position.

I   6   Independent variable
A factor that influences another factor known as the dependent variable (D9).

I   7   Index numbers
A statistical series denoting relative values, usually where there is no absolute measure, and almost invariably standardized to one item of the series given as 100. The time or place given as equalling 100 is called the base. Separate articles will be found under the following headings:

C11   Chain-base index number
F4   Fisher's ideal index number
F6   Fixed-base index number
L4   Laspeyres's index number
P1   Paasche's index number

to which reference should be made for definitions and formulae. A few other entries are more specific:

C30   Cost-of-living index (retail-price index)
P25   Price index
P33   Productivity
V2   Value index
V3   Value of money
V13   Volume index

Some 200 or 300 index number formulae have been propounded over the years — some the same in different guises, many impracticable for one reason or another, some erratic, a few heavily biased — and all can be described either as basically arithmetic or basically geometric in form, though the former includes a type of formula generally referred to as being of aggregative form.

Irving Fisher, famous for his 'ideal' formula (F4), advocated the adoption of formulae of the Laspeyres-Paasche group in his standard work, *The Making of Index Numbers,* first published in 1922, but even so, geometric formulae remained much in favour despite their bias and numerous complexities: the Board of Trade, for instance, continued compiling its index of wholesale prices on the basis of a geometric formula until 1953, when it abandoned it in favour of a series of separate index numbers, all based on an arithmetic formula — that of Laspeyres, to be precise. But that only followed what was becoming universal custom, under pressure from such international bodies as the Organization for European Economic Cooperation and the Statistical Section of the United Nations Organization.

The above applies only to index numbers of prices. In general, quantity index numbers, such as the index of production, are calculated by dividing the value index — the total value of production in terms of money, standardized to a base year = 100 — by the corresponding price index. If the price index is Laspeyres, then oddly enough, the production index is Paasche. Let $V_{n \cdot 0}$ be the value index for the given period, n, to the base period, 0. Then, to the base 1.00,

$$V_{n.0} = \frac{\Sigma(p_n q_n)}{\Sigma(p_0 q_0)} \tag{7.1}$$

where p is the price of the individual items in the period, and q is the quantity produced. Dividing this by Laspeyres price index, we have

$$Q_{n.0} = \frac{\Sigma(p_n q_n)}{\Sigma(p_0 q_0)} \div \frac{\Sigma(p_n q_0)}{\Sigma(p_0 q_0)}$$

$$= \frac{\Sigma(p_n q_n)}{\Sigma(p_n q_0)} \tag{7.2}$$

which is Paasche's quantity index.

There is more evidence of the close relationship between Laspeyres and Paasche. Of the several tests of a good index number, one is called the time-reversal test, which makes the hypothesis that exchanging the subscripts n for 0, and 0 for n in the formula should provide the index number for the base period relative to the given period = 1.00:

$$\text{Laspeyres's price index thus time reversed} = \frac{\Sigma(p_0 q_n)}{\Sigma(p_n q_n)}$$

which is the inverse, not of Laspeyres's price index, but of Paasche's. Paasche's index treated in the same way gives the inverse of Laspeyres's.

Where, for a price index, all prices change in the same proportion from the base to the given period, there is no doubt at all that the correct price index is equal to the price relatives (P26), P_n/p_0, of any of the items. If all prices rise by 10%, then the price relatives are all equal to 1.10, and the index, too, is 1.10 to base = 1.00. Both Laspeyres and Paasche satisfy this simple test, ie they would give the same solution, $P_{n.0} = 1.10$. We have, then

$$\begin{array}{ccc} Laspeyres & = & Paasche \\ \\ P_{n.0} \quad = \quad \dfrac{\Sigma(p_n q_0)}{\Sigma(p_0 q_0)} & = & \dfrac{\Sigma(p_n q_n)}{\Sigma(p_0 q_n)} \end{array}$$

Therefore
$$\frac{\Sigma(p_0 q_n)}{\Sigma(p_0 q_0)} = \frac{\Sigma(p_n q_n)}{\Sigma(p_n q_0)}$$

each side of which is the corresponding quantity index number. It follows that, where either prices or quantities all change in the same proportion from the base to the given period, Laspeyres and Paasche give the same solution for both the price and quantity index numbers, but what is more important, we know this to be the correct solution. In normal conditions, where neither prices nor quantities often change in the same proportion, a solution can be described as correct only on the criterion of the formula employed. There is little doubt, however, that the true index number lies somewhere between the Laspeyres and Paasche solutions.

So far, no weighted index-number formula has yet been pro-pounded to satisfy what has come to be called the *circular test,* which makes the hypothesis that, of a continuous series of index numbers, it should be logically possible to compare any two. Strictly, any weighted index number of a continuous series can be compared with the base only. In practice, however, comparisons with other numbers of the series occur daily, and there is no evidence that it results in serious harm. It happens sometimes that a non-continuous series of precisely weighted index numbers is required for a special *ad hoc* analysis of some kind. The best procedure, then, is to use accumulated total values for each item in all periods or places for weighting purposes. For this, the aggregative type of formulae, as used in the above arguments, cannot be applied where period or place provides the base. It is necessary to employ the arithmetic type; the price index of the closed system would be

$$P_{n.0} = \frac{\Sigma\left(v_a \frac{p_n}{p_0}\right)}{\Sigma(v_a)} \tag{7.3}$$

where v represents value, and subscript a denotes that v is the accumulated value.

Since $v_a = p_a q_a$, where p_a is the average price of the item in all periods or places and q_a is the accumulated quantity of the item, an alternative to (7.3) is to use the average of all periods or places as the base = 1.00. For a price index, we would then have

$$P_{n.a} = \frac{\Sigma\left(p_a q_a \dfrac{p_n}{p_a}\right)}{\Sigma(v_a)} \qquad (7.4)$$

which, as p_a in the numerator cancels out, can be reduced to the aggregative form:

$$P_{n.a} = \frac{\Sigma(p_n q_a)}{\Sigma(v_a)} \qquad (7.5)$$

On a more practical plane, it can be said that the weighting system is all important. Equations (7.3) and (7.4) make it clear that the weights are the values, not the quantities in a price index, or the prices in a quantity index. Since value statistics are much more readily available than quantity statistics, the arithmetic form of Laspeyres's price index is more practicable than the aggregative form. It is largely because quantity statistics are hard to come by that quantity index numbers are usually made by dividing the value index by the price index.

In practice, price data for a price index usually comprise a sample of prices within the orbit of the inquiry. The items covered are usually grouped, and a base-period value weight is allotted to each group, independently of the value of the items forming the sample. A subindex for each group is then made, and the subindex numbers are brought together, each weighted by reference to the allotted base-period weight, to form the main index. Laspeyres's arithmetic formula is used for this, the subindex numbers taking the place of the price relatives in the numerator.

Quantity index numbers for some purposes such as productivity measurement are best weighted by reference to labour content rather than value. See P33; also M4, in which a 'connoted' marginal cost is used in lieu of price in a production index.

A measure of the sampling error (S4) can be obtained by taking the standard deviation of the relatives, the square of whose mean deviations are weighted by reference to the pattern of value weights used in the construction of the index. A numerical example will make this clearer. Table I7.1 shows how a price index is made, and the calculation of the standard deviation. If the example applies to a group of items making a subindex number, the subindex numbers are gathered together in the same way, with the allotted group weights used for weighting; the sampling error of the combined index number is thus calculated.

Applications.   Stock valuation, productivity index making etc.

Table I   7.1   Laspeyres's price index and sampling error.

Item	Value weight $(p_0 q_0)$	Base period p_0	Price given period p_n	Relative p_n/p_0	Relative weighted $p_0 q_0 \frac{p_n}{p_0}$	Deviation of relative from mean $\frac{p_n}{p_0} - 1.25$	Deviation squared Actual	Weighted
	£'000	£	£					
A	1	2	3	1.50	1.50	0.25	0.0625	0.0625
B	4	5	5	1.00	4.00	−0.25	0.0625	0.2500
C	10	2	4	2.00	20.00	0.75	0.5625	5.6250
D	2	4	5	1.25	2.50	0	0	0
E	3	8	4	0.50	1.50	−0.75	0.5625	1.6875
Total	20			6.25	29.50	0		7.6250
Average				1.25				

Weighted average (÷ 20)   *1.475* (to base = 1.00)   0.38125

                                            or   *147.5* to base = 100

The standard deviation is $\sqrt{(0.38125)}$ = 0.6174

And the sampling error is $\dfrac{0.6174}{\sqrt{(5-1)}}$ = 0.3087

I   8   Induction

Sometimes called *a posteriori* reasoning, working back from effect to cause; also, arguing from the particular to the general, or attempting to establish general principles from observed individual cases. For a more detailed discussion, see deduction (D4).

Applications.   Statistical research, correlation analysis, model building.

I   9   Input-output

A generic term for the process of comparing and analysing input and output in respect of an individual person, a group of people working as a team, a company or group of companies, and the industry of the country. At the individual level, the process is generally referred to as

Interest rates have increased manifold since 1932-51, when they stood at record low levels. The Treasury bill rate in mid- 1951 stood at about 0.5% (per annum), and had risen to 7.8% by mid-1969. Medium- and long-dated government bonds had increased from about 3.5% in 1951 to more than 9.0% by mid-1969. There has been little, if any decline since 1969; indeed, the tendency has been for interest rates to continue rising. The prevailing high rates make interest on borrowed money a business cost to be reckoned with. For the various applications of the rate of interest in compounding and discounting, see A8, 9, 11, 12 and 13, C3, 16 and 29, D5 and 16, F12, I11, N4, P23, S18, T3, U4.

I   11   Interest tables

There are many published volumes of interest tables. Most are concerned with short-period loans such as are made by private moneylenders and banks, necessarily at simple interest. For business statistics and management-accounting purposes, annual compound-interest tables are more useful. One of the best-known is *Parry's Valuation Tables* (Estates Gazette Ltd), which provide all the basic series, mostly for one to 100 years and for a wide range of interest rates. Dual-rate figures are given for the present value of £ a year (years' purchase), and dual-rate and after-income-tax figures are also given for the present value of £ a year. The highest sinking-fund rate included in the dual rate seems much too low in these days of high interest rates.

I   12   Interfirm comparison

Part of the work of the Centre of Interfirm Comparison Ltd, an organization founded in 1959 by the British Institute of Management and the British Productivity Council. Comparison is made by reference to certain business ratios (B17) of participating firms, which are provided with reports of the results of the comparison of their own ratios with those of their competitors; no confidential information is given away, however.

The major business ratios form a pyramid, the apex consisting of:

$$1 \quad \frac{\text{Operating profit}}{\text{Operating assets}}$$

followed below by three ratios:

$$2\ \frac{\text{Operating profit}}{\text{Sales}} \qquad 3\ \frac{\text{Sales}}{\text{Operating assets}} \qquad 3a\ \frac{\text{Operating assets}}{\text{Sales}}$$

Ratios 2 and 3a are then followed by a number of ratios, all with sales or the sales value of production as the denominator. The numerators of these ratios are as follows:

4   Production cost of sales
5   Distribution and marketing costs
6   General and administrative costs
7   Cost of materials
8   Other production costs
9   Works labour cost
10 Current assets
11 Fixed assets
12 Material stocks
13 Work in progress
14 Finished stocks
15 Debtors
16 Land and buildings
17 Plant and machinery
18 Vehicles

Ratios 7, 8 and 9 have *sales value of production* (C31) in the denominator, all the others have *sales*. It will be seen that ratios 12, 13, 14 and 15 add up to ratio 10, and that ratios 16, 17 and 18 add up to ratio 11.

The above are the major ratios appropriate to a manufacturing company; certain minor ratios are also provided, depending on the industry, to throw light on causes of differences or changes in certain of the major ratios.

The address of the Centre for Interfirm Comparison is the same as that of the BIM: Management House, Parker Street, London, WC2.

I   13   Intermediate products

Goods at an intermediate stage of production—neither raw material as purchased by a manufacturing firm, nor the finished article ready for sale to customers; an important concept in the censuses of production (C9). The output section of the detailed census forms is largely devoted to sales, but for some intermediate products, the quantity produced, called the *total make,* is asked for as well. It should be noted that the total make of intermediate products, such as iron castings, includes not only the quantity of the product sold as such, but also the quantity used in the same factory in further processing and in assembly. Statistics of total make are published in the Census of Production reports.

I   14   Internal rate of return (IRR)

That rate of discount which converts the total yield — gross of depreciation — of a capital project, to a present value (P23) equal to the capital outlay (C4) on the project. Of the three principal discounted-cash-flow techniques that have been propounded since the early sixties, it is the one that has been the most misunderstood by its proponents, as well as by its critics. Some critics have argued that the IRR technique assumes that the whole of the yield goes into a depreciation sinking fund (S18) to accumulate at a rate of interest equal to the IRR discount rate. And some proponents of the IRR reply by stating that no sinking fund of any kind is involved in, or implied by, the IRR technique. It seems that both are wrong. As an understanding of the true position in this respect is fundamental to an understanding of the IRR technique itself, it is worthwhile going into the matter fairly deeply. The relevant discounting principle may be stated as follows: *the annual yield of £x of capital invested now for* n *years at* r *interest per £ is equal to the annual sinking fund to provide £x at the end of* n *years at* r *interest per £, plus the rate of interest,* r *per £.*

Suppose a fixed asset (F5) costing £4000 is expected to yield £1000 a year during its life of 15 years; the IRR would be 24%, ie the present value of £1000 a year for 15 years invested at 24% interest is £4000:

$$£1000 \text{ at } £4.00 \text{ per } £ = £4000$$

(The figure of £4.00 per £ is taken from a set of published compound-interest tables). Adopting 24% as the rate of interest, then applying the principle stated above, we have:

Annual sinking fund at 24% to give £4000 in 15 years:

	£
Depreciation £4000 at £0.10 per £	40
Interest on £4000 at 24%	960
Total, the annual yield of the asset	£1000

(Published sinking-fund tables do not give figures for interest rates higher than 20%. The figure of £0.010 per £ has been calculated from the formula given in S18).

But, say the critics, you cannot do that. You are using the IRR as a rate of interest, and it is not a rate of interest, and they quote Keynes in support of their argument: 'We must ascertain the rate of interest from some other source, and only then can we value the asset by "capitalising" its prospective yield'. (General Theory, p. 137.) Compare the above principle of discounting and analysis with the annual-value technique briefly discussed under A12. For other criteria and techniques, see N4, T3 and 15.

I   15   Interpolation
The process of estimating the value of a dependent variable, for some given value of an independent variable that lies within an observed range, from a graph or regression or similar equation. An example is the estimation of the population of a town in, say, 1966 from the census figures of 1961 and 1971. See G2 for such an example; for extrapolation, see E12.

I   16   Interquartile range
The difference between the upper and lower quartiles (Q5). see dispersion (D19).

I   17   Investment criteria
The several techniques of converting the yield of an asset to a capital sum, or the capital outlay to an annual cost, so that they can be added together or deducted one from the other; also the several different ways of expressing the net yield, more precisely called the yield criteria (Y3). For the techniques, see A12, I14, N4, T3 and 15.

I   18   Invisible trade
That part of the country's trade with other countries which consists

of capital transfers and services, eg banking, insurance, shipping, interest, profit and dividends, travel, aviation, private capital transfers and government transactions. There is always a substantial favourable balance of trade on invisible account, and it has grown appreciably in recent years. Table I18.1 shows the credits, debits and net balance for all invisibles for 1963-72.

Table I   *18.1*   *UK invisible trade, £ million.*

	Credits	Debits	Balance, net credit
1963	2 482	2 278	204
1964	2 646	2 509	137
1965	2 871	2 683	188
1966	2 953	2 796	157
1967	3 239	2 997	242
1968	3 803	3 426	377
1969	4 300	3 713	587
1970	4 906	4 237	669
1971	5 430	4 689	741
1972	5 851	5 141	710

Source: Financial Statistics, HMSO.

j J

J1   Joint cost

The cost of inputs that are inseparably attributable to two or more products; the indivisible part of the cost of joint products (J4), such as coal-gas, gas-coke and sulphate of ammonia; beefsteak and sirloin; petrol and diesel oil.

J2   Joint demand

Said of products and services that are in demand one with another. Petrol for motor vehicles, lubricating oils and tyres are said to be in joint demand; so, too, are bread and butter. The joint demand for petrol, oil and tyres for cars is derived from the demand for travel, which is described as being in joint demand with petrol, etc. Petrol, etc. is in an indirect joint demand for travel. The demand for the factors of production of a consumer good is an indirect joint demand for the consumer good. And if one of the factors consists of a raw material, the factors of production of that raw material are an indirect joint demand for the material. Marshall, in his *Principles of Economics,* refers to consumer goods as being of the first order; to the materials and equipment used in the production of consumer goods as producers' goods, as of the second order; to the goods and equipment used by the producer of second-order goods as of the third order; and so on.

Applications.   Demand-model building, especially for producers' goods.

J3   Joint function

A function (F11) in which the effect on the dependent variable (D9) of a change in one independent variable (I6) is influenced by the level of another independent variable. Joint functions may be rare in manufacturing industry; but the mathematical model builder (M2) can scarcely afford to ignore their potential existence. M. Ezekiel, in his *Methods of Correlation Analysis* (1930), instanced the yield of corn as a function of rainfall and temperature, and came to the conclusion from the relevant statistical evidence available that the effect on yield of an increase of one inch of rainfall varies with temperature. The yield of corn is thus shown to be a joint function of rainfall and temperature. There is another point about this example worthy of the attention of the mathematical model builder: that the effect on yield of a change of one inch in the rainfall during the growing season, or of a change of one degree in the temperature, necessarily varies with the average temperature, or the total rainfall, during the season. Very high rainfalls or average temperatures may have an adverse effect on yield, just as very low rainfalls or temperatures have. In short, there is an optimum total rainfall and an optimum average temperature. Ezekiel showed that the optimum total rainfall in June, July and August in 1927 in the area of the Corn Belt covered by his statistics was 11 inches, and the optimum average temperature for the same months was 74°F.

A solution to a problem involving optimum values of two or more independent variables calls for a large number of sets of observations, and the shape of the mathematical model would necessarily have to be empirically determined. In a three-variables problem, the effect of changes in one of the two independent variables, X_3, would need to be eliminated by grouping all sets of observations in such a way that X_3 would be represented by a series of relatively narrow bands. Then the observed values of the dependent variable, X_1, and the other independent variable, X_2, for each group would be plotted on rectilinear graph paper.

Suppose the observed values of X_1 and X_2 are as follows for a group range of X_3 of 15-18, with a group average of 16.7:

$X_1 = $ 7.5 8.4 9.1 10.0 10.1 9.9 9.6 9.4 7.4
$X_2 = $ 10.1 11.8 14.2 17.5 18.0 21.0 23.0 25.2 30.0

When plotted, a smooth freehand graph fitted to the scatter of dots has the appearance of a quadratic. Proceed to test for a quadratic by applying the method of finite differences as demonstrated in the article under F3. It will be found that the appropriate empirical equation is

$$X_1 = X_2 - 0.025X_2{}^2 \text{ (for } X_3 = 16.7)$$

By equating dx_1/dx_2 to zero, it will be found that the maximum value of X_1 is 10, when $X_2 = 20$ (and $X_3 = 16.7$)

This process is repeated for the observed values of X_1 and X_2 corresponding to the next group range of X_3, which may be 18-21, with an average of 19.4, and so on for each group range of X_3. Finally, a two-dimensional table showing the estimated values of X_1 for a range of values of X_2 and of X_3 may be compiled, as in Table J3.1.

Table J3.1   Estimates of X_1 for different values of X_2 and X_3.

Average values of X_3	...	16.7	19.4	22.5	25.6	...
Values of X_2						
10		7.5				
12		8.4				
14		9.1	...			
16		9.6				
...		...				

If X_1 is a linear function of X_3, the coefficients of X_2 and $X_2{}^2$ will turn out to be much the same for all equations expressing X_1 as a function of X_2 for the different average values of X_3. In that event, there would be different values for the unattached constant, which would equal $a_3X_3 - 16.7a_3$ where a_3 is the coefficient of X_3 in the complete equation

$$X_1 = X_2 - 0.25X_2{}^2 + (a_3X_3 - 16.7a_3)$$

The maximum value of X_1 would be found in the first column, Table J3.1 where a_3 has a minus sign, and in the last column, where it has a plus sign. But this is unlikely to be the case with the conditions postulated. Some general reasoning may help. For instance, it could well be that where X_2 is zero, X_1 must also necessarily be zero, in which case none of the equations could have an unattached constant, X_1 would be a curvilinear function of X_3, and the coefficients of X_2 and $X_2{}^2$ would vary with changes in the value of X_3.

J4 Joint products

Products that are made or won necessarily simultaneously; usually with one as the main product and the other, or others, as by-products. In some cases the by-products originated as waste and were disposed of as such. Brewers' yeast is an example. Before the nutrient properties of yeast were discovered, it was disposed of as waste. Now human- and cattle-food manufacturers purchase it from the breweries. Coal-gas works have many by-products: coke, crude coal-tar, processed coal-tar yielding creosote oils, pitch and naphthalene, sulphate of ammonia, crude benzole, benzole and refined tar. Oddly enough, there is another but less well-known industry — coke ovens — whose main product is coke, the by-products consisting of coal-gas and the same by-products, apart from coke, as those of the coal-gas industry.

It will be observed that the gas and coke industries process some of their by-products. They sell part of some in their crude state, the other part being processed prior to sale. The economics of by-products and waste is a somewhat complex subject. It might happen that the cost of processing a by-product exceeds the sales proceeds from it, but if the cost of disposing of it as waste exceeded the resulting loss, it would pay to process it and thus convert it into a saleable commodity. The loss would form part of the cost of the main product, just as would the total cost of disposing of the by-product as waste.

It may well be that, logically, the total cost of processing all by-products should be charged to the main product, and all sales proceeds from by-products credited to the main product. It is a difficult problem. In any event, it seems that any attempt to allocate the joint costs (J1) of manufacturing the main products and by-products would serve no useful purpose. It would involve the marginal cost (M4), as well as the overheads. Where two or more products are genuinely joint, if, for instance, a gas works or a coke oven were in business as much to produce and sell coke or gas as to produce and sell gas or coke, there may be a case for cost allocation, even though it may involve the marginal cost. How then, would one decide a rational pricing policy? It would seem that the correct approach is to calculate the marginal cost of each product from the total joint variable manufacturing cost, and not to allocate it. Any further marginal cost subsequently incurred in separate processing would be added to the initial marginal cost.

Joint products do not include goods made simultaneously for economic reasons, or merely because they are sold together, such as teacups and saucers, and table knives and forks. See A3 and 15, R6, W1.

k

K1 Key statistics
A short name for the publication *The British Economy: Key Statistics 1900-1964.* The London and Cambridge Economic Service (Times Publishing Co.).

Applications. Market research.

K2 KS statistics
Said to be short for *keep-in-step statistics,* introduced during World War II for economic control purposes, but found to be so useful to industry and the Government that their collection and collation have been continued. They consist, for the most part, of sales and delivery figures for a wide variety of products, on a monthly or quarterly basis. The figures are published in the *Monthly Digest of Statistics, The Annual Abstract of Statistics* and the Department of Trade's *Business Monitor.*

Applications. Market research.

1

L1   Labour-time function

Man hours or man days expressed as a function of work done; usually of linear form, and not to be confused with the employment function (E5), which is of logarithmic form. A worked example of a regression analysis by cross-classification of a labour-time problem is given in C36, and by least squares in L7.

It is not possible to analyse by any regression method the average time spent on different processes in the manufacture of an article, when the number of times one process is carried out is the same as, or a multiple of, that of any other process. Regressio ι analysis is in effect the solution of simultaneous equations, and therefore its successful application depends on internal variations. In a problem in three variables, for instance, in which process A is performed twice as often as process B, the number of times that process A is carried out by one man in a week would always be double that of process B. Ignoring any unattached constant, we might have two equations, such as

$$100a + 50b = 40 \text{ (man hours)}$$
$$120a + 60b = 50 \text{ (man hours)}$$

where a and b are the coefficients of A and B in the equation, ie the time spent on the two processes in the week:

$$aA + bB + K = \text{weekly man hours}$$

Any attempt to solve these equations simultaneously would give

$$0a + 0b = 2 \text{ man hours}$$

which is nonsense, though it does indicate that the work is done more efficiently in a 40-hour week than in a 50-hour week.

L2   Labour turnover

The ratio of the number of personnel leaving the firm, in unit time, to the total number employed. Unit time is usually a calendar month, and the number employed is the average during the month, the average being, say, that of the number employed on the first and last working days and the working day nearest the middle of the month.

Another measure of labour turnover is the average length of stay with the firm. If the ratio for a month turns out to be 0.05, ie 1/20, the average length of stay is 20 months. The smaller the ratio, the longer the average length of stay: a ratio of 0.025 is equal to 1/40, or an average stay of 40 months.

Labour turnover for the firm can be regarded as a key business ratio (B17) and analysed in a number of ways: by reference to (1) age and sex; (2) reason for leaving; (3) salary grades and wages grades; (4) departments and establishments. Analysis by (1) or (2) would be additive, but (3) and (4) would more logically give each constituent its own denominator, and ratios would therefore be non-additive.

The lower the ratio, the better: it belongs to that kind of ratio which the firm likes to see falling. The so-called learning law discussed under employment function (E5) indicates the chief advantage of minimizing the labour turnover, and thus maximizing the average length of stay.

Applications.   Personnel management.

L3   Lag in time

A time lag between cause and effect. Time lags are difficult to detect and more difficult to measure. They are no respecters of statistical units of time. It is always unlikely that a time lag will consist of an exact unit of time or an exact number of units of time. The shorter the unit of time, the easier it is to detect and measure, and therefore the easier it is to advance the statistical series of the dependent variable to allow for the lag. Some time lags can be eliminated by synthesis ie by building up one figure or series of figures to conform to the period of the other. Cost accountants, for instance, eliminate the time lag between input and output by building up a cost of sales figure (C31) to compare with the figure of sales proceeds. From the statistician's viewpoint, synthesis is not a satisfactory way of compiling a figure of any kind. It assumes away random variations, and thus may assume away the measure of the phenomenon he is seeking to isolate.

In practice, time lags vary in duration from one period to the next. The lag between input and output depends on whether output is being increased or run down, and whether material stocks are being built up or run down. The best that can be done by statistical analysis is to achieve a solution consisting of an average time lag to the nearest whole unit of time. It is done by calculating the coefficient of correlation (C25), first without adjusting for a time lag, and then, after adjusting for each possible successive time lag, until the coefficient of correlation begins to decline. The adjustment giving the maximum coefficient provides an estimate of the average time lag to the nearest whole unit of time.

Another difficulty that may arise in practice is that a change in an independent variable may take time to achieve its full effect on the dependent variable. Advertising is an example. Point-of-sale advertising may take less time to exert its full effect on sales than will newspaper or TV advertising, but whatever media are used, a campaign takes time to come to full fruition.

Finally, the problem of time lags becomes almost intractable where a multiplicity of independent variables has to be considered. If each of m independent variables is to be lagged for analytical purposes for each of 0 to n periods, the number of multiple-correlation coefficients that would need to be calculated to determine the lag for each independent variable would be $(n + 1)^m$. For three independent variables each lagged up to two periods, the number of coefficients to be calculated would be
$$(2 + 1)^3 = 27$$
What is more, it would first be necessary, as shown under C25, to determine the form of the basic mathematical model. This would have to be done by general reasoning alone, since it would not be reasonably possible to test the model empirically until the basic data had been adjusted for any time lags.

Applications.   Regression and correlation analysis.

L4   Laspeyres's index number
The formula that is now the most widely used throughout the world

for making price-index numbers. The fixed-base formula (F6) is

$$P_{no} = \frac{\Sigma\left(p_0 q_0 \frac{p_n}{p_0}\right)}{\Sigma(p_0 q_0)} \tag{4.1}$$

where $P_{n.0}$ is the price index number for period n to the base period $0 = 1.00$, p_n/p_0 Σ the sign of summation, and p and q the price and quantity sold. The ratio p_n/p_0 is called the price relative (P26), and since p_0 cancels out, the formula reduces to

$$P_{n.0} = \frac{\Sigma(p_n q_0)}{\Sigma p_0 q_0} \tag{4.2}$$

which is generally referred to as Laspeyres's aggregative formula, as distinct from his arithmetic formula in (4.1). Although it seems more complex, the arithmetic formula is the more practicable of the two. It neccessitates calculating the price relatives, but saves calculating $p_n q_0$ for each item. It also shows more clearly the weighting system, the weights consisting of the values pq in the base period for each item.

Of all weighted index-number formulae, Laspeyres's is the simplest, both to understand and to apply. It is based on the basket-of-goods principle. If a basket of goods costs £5.00 in the base period, and the same basket costs £5.50 in the given period, the price index in the given period to the base is

$$\frac{5.50}{5.00} = 1.10$$

The simple example in Table L4.1 demonstrates the application of both the arithmetic and aggregative forms of Laspeyres.

Quantity index numbers are usually derived by dividing the price index into the value index:

$$V_{n.0} = \frac{\Sigma(p_n q_n)}{\Sigma(p_0 q_0)} \tag{4.3}$$

where $V_{n.0}$ is the value index to base = 1.00. This produces an index weighted by values in the given period, called Paasche's index number (P1). However, Laspeyres's aggregative quantity index is given below, for what it is worth.

$$Q_{n.o} = \frac{\Sigma(p_0 q_n)}{\Sigma(p_0 q_0)} \qquad (4.4)$$

Table L4.1   *Applying Laspeyres's index number.*

Item	Quantity in base	Price in base	Value in base	Price in given period		$\frac{p_n}{p_0}$	$p_0 q_0 \frac{p_n}{p_0}$
	q_0	p_0	$p_0 q_0$	p_n	$p_n q_n$		
A	6	10	60	11	66	1.10	66
B	5	12	60	15	75	1.25	75
C	2	8	16	10	20	1.25	20
Totals			136		161		161

$$P_{n.0} = \frac{161}{136} = 1.18 \text{ to base} = 1.00$$

For a discussion of the relationship between Laspeyres and Paasche and of index numbers in general, and for a full list of references, see index numbers (I7).

Applications.   Stock valuation; calculating trends in materials prices and wage rates to facilitate product costing and forecasting.

L5   Law of error
States that the frequencies with which errors of measurement and differences between actual figures and estimates occur tend to form a symmetrical distribution curve which, in terms of standard deviations (S24), conforms to the normal curve of error (N9); sometimes called the *normal probability curve* or the *normal frequency-distribution curve.* See normal curve of error (N9).

L6   Learning curve
A curve which describes the upward trend in productivity, or the downward trend in labour requirements, per unit of output as the employees gain experience in carrying out particular processes or in producing certain goods. The theory underlying the concept is that the longer a man is engaged in performing the same task, the more

proficient he becomes and the greater his rate of output. The problem of the statistical measure of the effect of learning on productivity is discussed under the employment function (E5).

L7   Least squares

The most sophisticated but, nevertheless, probably the best method of regression analysis (R7). Provided there is no intercorrelation (C7 and 25), it conforms to the theory of error, and it is therefore possible to calculate the standard errors (S25) of the regression coefficients (R8) and the standard error of estimate of the regression equation (R9).

Table L7.1 gives the least-squares formulae, and shows how they can be built up, for any number of variables. The basic equation may be linear or logarithmic:

Linear: $X_1 = K + a_2 X_2 + a_3 X_3 + \ldots a_n X_n$ (7.1)

Logarithmic: $\log X_1 = K + a_2 \log X_2 + a_3 \log X_3 + \ldots + a_n \log X_n$ (7.2)

Table L7.1   Least-squares formulae for any number of variables, and how to build them up.

Equation no.		X_2		X_3		....	X_n		X_1
(1)	X_2	$\Sigma(x_2{}^2)a_2$	$+$	$\Sigma(x_2x_3)a_3$	$+$	....	$\Sigma(x_2x_n)a_n$	$=$	$\Sigma(x_2x_1)$
(2)	X_3	$\Sigma(x_2x_3)a_2$	$+$	$\Sigma(x_3{}^2)a_3$	$+$	....	$\Sigma(x_3x_n)a_n$	$=$	$\Sigma(x_3x_1)$
...	...			...			...		...
(n−1)	X_n	$\Sigma(x_2x_n)a_2$	$+$	$\Sigma(x_3x_n)a_3$	$+$	....	$\Sigma(x_n{}^2)a_n$	$=$	$\Sigma(x_nx_1)$
(n)	X_1	$\Sigma(x_2x_1)a_2$	$+$	$\Sigma(x_3x_1)a_3$	$+$	....	$\Sigma(x_nx_1)a_n$	$=$	$\Sigma(x_1{}^2)$

Note: Σ is the sign of summation.

The linear form − (7.1) − may be used for second-order equations and upwards, in which case X is represented by X_2, X^2 by X_3, and so on; however, it is rarely advisable in practice to go higher than X^3, since it is theoretically possible to obtain a perfect fit − to find an equation of some higher degree that would satisfy any number of sets of observations and thus accommodate all random variations (R2). Thus, three sets of observations might need a quadratic, four sets a cubic, five sets, a fourth-degree equation, and so on, always supposing that an equation of lower degree would not fit. See quadratics (Q2).

In the practice of least squares, the deviations from the mean of each series are used as basic data. The deviations are represented by x_1, x_2, x_3, and so on; these are shown in Table L7.1. Mean deviations are used for technical reasons, the principal one being that they reduce the working number of significant figures necessary to give reasonably accurate results; they also have the effect of eliminating the unattached constant shown as K in (7.1) and (7.2) above. To fill in the term in each cell of the table, the symbol in the column head is multiplied by that in the stub, and the coefficients are given the subscripts of the symbols in the column heads. Each line represents an equation for simultaneous solution with others where more than two variables are involved.

For two variables, we take the relevant part of the first equation for one solution, and of the last equation for an alternative solution. The former solution is called the regression of Y on X, and the latter, the regression of X on Y. Strictly, in the notation of Table L7.1, they are X_1 on X_2 and X_2 on X_1.

Regression of Y on X
(or X_1 on X_2)

$$\Sigma(x_1{}^2)a_2 = \Sigma(x_2 x_1)$$

Therefore $\quad a_2 = \dfrac{\Sigma(x_2 x_1)}{\Sigma(x_1{}^2)}$ $\hfill$ (7.3)

Regression of X on Y
(or X_2 on X_1)

$$\Sigma(x_2 x_1)a_2 = \Sigma(x_1{}^2)$$

Therefore $\quad a_2 = \dfrac{\Sigma(x_1{}^2)}{\Sigma(x_2 x_1)}$ $\hfill$ (7.4)

For a problem in three variables two equations for simultaneous solution are necessary, but we have three in Table L7.1 to choose from:

A $\qquad \Sigma(x_2{}^2)a_2 + \Sigma(x_2 x_3)a_3 = \Sigma(x_2 x_1)$
B $\qquad \Sigma(x_2 x_3)a_2 + \Sigma(x_3{}^2)a_3 = \Sigma(x_3 x_1)$ $\hfill$ (7.5)
C $\qquad \Sigma(x_2 x_1)a_2 + \Sigma(x_3 x_1)a_3 = \Sigma(x_1)^2$

These provide three sets, each of two equations: A and B, B and C, and A and C, so that there are three solutions for each of the two coefficients a_2 and a_3. They are called the regressions of X_1 on X_2, X_3; of X_2 on X_1, X_3; and of X_3 on X_1, X_2, respectively.

Similarly, for a problem in four variables three equations are necessary for simultaneous solution, and there are four to choose from, so that each coefficient has four solutions; with a problem in five variables, each coefficient has five solutions; and with a problem in n variables, each coefficient has n solutions.

Returning to the problem in three variables, we have equations A and B giving the regression of X_1 on X_2, X_3. How do we tell that it is A and B that give the regression of X_1 on the others? It will be seen that each of the equations A, B and C contains, on one side or the other, the sum of squares of mean deviations: equation A has $x_2{}^2$, B has $x_3{}^2$ and C has $x_1{}^2$. The square omitted from the two equations giving a solution indicates the variable whose regression it is on the other variables. Equations A and B do not contain $x_1{}^2$, so that they give the regression of X_1 on X_2, X_3. This may all seem somewhat academic, but in both industry and economics it is of some practical importance: the two or more solutions have different and definable meanings. The regression of X_1 on the rest assumes that X_1 is the true dependent variable, and X_2, etc. are true independent variables, and that there is no interdependence between X_1 and X_2, between X_1 and X_3 and so on.

In the field of industry, we know that clear-cut dependence and independence rarely exist, so that some kind of average of solution values may have to be struck – a process which statistical theorists used to frown upon, and may still do. Where there is interdependence or *intercorrelation*, as it is generally called, the analyst may consider adopting an average of the solution values of each coefficient, weighted by reference to a subjective estimate of the interdependence each variable has with each of the others. For a problem in two variables, there is no difficulty: it may be thought, for example, that X_1 depends on X_2 twice as much as X_2 depends on X_1; then if the two solutions of a_2 are 20 for X_1 on X_2, and 30 for X_2 on X_1, the weighted average would be

	20 x 2	40
	30 x 1	30
Total	3	70
Weighted average		23.33

For problems in three or more variables, there is scope for the analyst to exercise his ingenuity, as well as his powers of reasoning from cause to effect, for it is necessary to keep in mind the possibility that the variables adopted as the independents for purposes of the regression analysis may be interdependent among themselves, as well as with the variable adopted as the dependent variable.

As the literature contains nothing on the subject, the following is offered as an approach to, if not a complete solution of, the problem. A numerical example will best demonstrate the approach, the problem being, say, one in four variables in which one, X_4, is regarded as completely independent. A table of estimated dependencies is drawn up on the lines of Table L7.2. To standardize the measure of the dependencies, they are related to the total of 100, so that each figure is, in effect, a percentage.

Table L7.2   Estimated dependencies for weighting purposes.

Dependence of Dependence on	X_1	X_2	X_3	X_4	Total
X_1	–	3	7	–	10
X_2	11	–	6	–	17
X_3	5	9	–	–	14
X_4	30	15	14	–	59
Total	46	27	27	–	100

We now have to allocate these weights over three coefficients, a_2, a_3 and a_4, for each of which we have four values so that we require 12 weights. Since no variable can depend on itself in any measure, out of the total of 16 cells in the table, only 12 are effective. Allocation seems to be a difficult problem, but general reasoning suggests the following weighting patterns:

Regression of	a_2	a_3	a_4
X_1 on the rest	11	5	30
X_2 on the rest	3	7	–
X_3 on the rest	9	6	–
X_4 on the rest	15	14	–
Total	38	32	30

The weighted arithmetic mean for the solution values of each coefficient is then calculated. If, for a_4, the solution values are 10, 15, 6, 2, respectively, then the weighted AM would be 10:

Solution values	Weights	Product
10	30	300
15	0	0
6	0	0
2	0	0
Total	30	300

Weighted average 300/30 = 10

The weighted average value of a_4 is therefore equal to its solution value for the regression of X_1 on the rest.

In considering the logic of this approach, we should keep in mind that X_1 has a unique position in the basic model: it is adopted as the dependent variable and does not have its own coefficient. We should also note that in regression analysis any variable, even the one represented by X_4 in the example, can be adopted to serve as the dependent variable. Choice turns on external convenience rather than on the internal facts, though other things being equal, one would choose the variable which has the greatest degree of dependence on the rest.

Most regression analyses involving more than three or four variables, and upwards of 50 sets of observations, are now carried out by computer. It is understood that computer manufacturers will supply users with ready-made programmes for regression problems in any number of variables and sets of observations. Where these numbers are small, the use of a computer would scarcely be worthwhile. There are many such problems in business management.

A numerical example can be used to demonstrate least squares as a desk job with the aid of a desk calculator, and the two short cuts that can be applied to save time. It is proposed to apply least squares to the problem in three variables detailed in Table C36.1. We can use the same notation, with x_2, x_3, x_1 representing the mean deviations. One short cut is to reduce each series by the smallest figure in the series, thus giving 0 in place of the lowest figure. The entry under C36 shows that the basic model is linear, the literal equation being

$$H = a_2 M + a_3 C + K$$

Table L7.3 gives the reduced series in the first three columns, and the squares and products in the following columns. It will be seen that the deviations from the mean are not extracted for the purpose, the sums of squares and products of deviations being obtained by deducting an easily ascertained correction sum. This is the second short cut.

The correction sums are equal to the corresponding squares or products of the totals of the reduced series divided by 15, the number of observations. For instance, the correction sum for $\Sigma(M^2)$ to give $\Sigma(x_2{}^2)$ is $805^2/15 = 43\,202$, and that for $\Sigma(MC)$ to give $\Sigma(x_2 x_3)$ is $805 \times 480/15 = 25\,760$. That $\Sigma(x_2 x_1) = \Sigma(x_3{}^2) = 3190$ is a pure coincidence.

Applying the formula of (7.5), we have

A $\qquad\qquad 7115a_2 - \quad 545a_3 = 3190$
B $\qquad\qquad -545a_2 + 3190a_3 = -65$
C $\qquad\qquad 3190a_2 - \quad 65a_3 = 1498$

Solving two at a time simultaneously, we then have the following solution values of a_2, a_3 and K:

Equations	a_2	a_3	K
A and B: regression of H on M, C	0.454	0.0571	9
B and C: regression of M on H, C	0.471	0.0602	7
A and C: regression of C on H, M	0.477	0.3780	−15
Cross-classification solution (C36)	0.468	0.0750	6.5

The solution reached by applying the methods of cross-classification (C36) and finite differences in tandem is given to show the similarity of the results. The value of K is determined by using the averages of M, C and H in the way demonstrated under C36. When reduced series are used, the averages are based on the totals of the reduced series, with the lowest figure added back after multiplying it by the number of observations in the series. This correction is shown at the foot of the first three columns of Table L7.3.

Note that the three least-squares solution values of a_2 are remarkably similar, whereas those of a_3 are similar for the regressions of H on M,C and M on H,C, but significantly different for the regression of C on H,M. If the transport contractor accepted all the work offered to him, weekly man hours, H, would be a true dependent variable, and in that case, the correct least-squares solution values of a_2, a_3 and K are those of the regression of H on M,C, and the standard errors of the three values may be calculated. See standard error (S25).

Table L7.3   Least squares applied to a problem in three variables.

Reduced series

M	C	H	M^2	MC	MH	C^2	CH	H^2
0	45	0	0	0	0	2 025	0	0
26	55	4	676	1 430	104	3 025	220	16
30	35	9	900	1 050	270	1 225	315	81
34	0	8	1 156	0	272	0	0	64
40	20	14	1 600	800	560	400	280	196
56	35	20	3 136	1 960	1 120	1 225	700	400
60	50	25	3 600	3 000	1 500	2 500	1 250	625
62	15	24	3 844	930	1 488	225	360	576
64	30	24	4 096	1 920	1 536	900	720	576
68	35	27	4 624	2 380	1 836	1 225	945	729
70	35	29	4 900	2 450	2 030	1 225	1 015	841
71	15	28	5 041	1 065	1 988	225	420	784
72	25	25	5 184	1 800	1 800	625	625	625
74	50	32	5 476	3 700	2 368	2 500	1 600	1 024
78	35	31	6 084	2 730	2 418	1 225	1 085	961
805	480	300	50 317	25 215	19 290	18 550	9 535	7 498
750	525	600						
1 555	1 005	900						
103.7	67	60						

Subtract
correction sum

			M^2	MC	MH	C^2	CH	H^2
			43 202	25 760	16 100	15 360	9 600	6 000
			7 115	− 545	3 190	3 190	− 65	1 498
			$\Sigma(x_2{}^2)$	$\Sigma(x_2 x_3)$	$\Sigma(x_2 x_1)$	$\Sigma(x_3{}^2)$	$\Sigma(x_3 x_1)$	$\Sigma(x_1{}^2)$

Applications.   Deriving regression equations in the analysis of demand, cost, employment and labour time, and in similar analyses in the field of management.

L8   Life tables

A generic term for published tables based on the latest census of population, and covering the following:

1   The present value of a life annuity of £ a year.
   (a)   For a single life.
   (b)   For the joint continuation of two lives.
   (c)   For the longer of two lives.
2   Simple premium to secure £ at death.
3   Annual premium to secure £ at death.
4   Mortality.
5   Expectation of life.
   Sources. Parry's Tables, Annual Abstract of Statistics.
   Applications.   Pension-fund calculations.

L9   Likelihood

Probability (P28); generally reflected in the literature in the phrase *maximum likelihood,* which applies to a method of maximizing the probability that an average derived from a sample will agree with the corresponding average of the universe.

L10   Linear programming

A series of methods originating in operational research based on the assumption that relationships between factors are of a linear (straight-line) type. The methods include break-even analysis (B14) and optimizing the product mix (P34).

L11   Linear-regression analysis

Regression analysis, both simple and multiple, based on a linear model (M16). A worked example is given under C36 and L7.

L12   Link relatives

Trend ratios (T16). For worked example, see L14.

L13   Log paper

There are two kinds of log paper: *log-log paper,* in which both axes have ratio scales, and *semi-log paper,* which has a vertical axis with a ratio scale, and a horizontal axis with a normal scale. A ratio scale is one on which any ratio is given equal distance on all sections of the scale. If the ratio 2 has two inches from 1 to 2, then the distance from 1½ to 3, or from 2 to 4, or from 10 to 20, will also be two inches. The absolute differences between the logarithms of each of these pairs of numbers is the same: 0.3010; the same applies to the differences between any other pairs of numbers that have the ratio. For a ratio of 1½, for instance, the absolute difference between the logarithms of the numerator and denominator is always 0.1761. To construct a ratio scale then, all one needs do is measure the required range of logarithms on a normal scale, insert the logarithms of whole numbers or appropriate fractions, at intervals, and the corresponding natural numbers, then erase the logarithms.

Plotted on semi-log paper, a geometric progression (G2) would describe a straight line, whereas on rectilinear graph paper, it would describe a curve. It used to be fairly common practice to plot time series on semi-log paper, even for presentation purposes and publication, but in recent years the practice has largely fallen into disuse. Presumably, the underlying assumption was that time series tend to rise or fall in geometric progression. Some, such as the GNP, undoubtedly do; others, such as company profits and share prices, undoubtedly do not.

Log-log paper is used by statistical analysts for plotting the statistical series of two factors, one against the other, where it is thought that there is a logarithmic relationship between the two. If the scatter of dots lies about a straight line, the hypothesis is probably true; if it lies about a curve, or is too wide to suggest a graph of any shape, the hypothesis is probably false. See correlation (C25); for the different shapes of graphs of natural numbers plotted against normal scales where they describe straight lines on log paper, see graphs (G6).

L14 Long-range planning

Corporate strategy based on the company's objectives and forecasts of sales, which provide the independent variables in equations for forecasting requirements of capital finance, manpower, fixed assets and materials. Planning ahead is the essence of long-range planning (sometimes referred to as LRP). The several dependent variables differ in the extent to which projections into the future need to go. Materials in manufacturing concerns call for the shortest of short-term sales forecasts, whereas the management development part of the manpower variable calls for the longest of long-term forecasts, probably at least 20 years in most companies. But it is generally recognized that forecasting a year ahead is hazardous enough, and that attempts at forecasting beyond five years are largely a waste of time. It is argued, on the other hand, that a long-term forecast attained by applying modern techniques to the best available data must be better than one reached by intuition. If the personnel manager needs a 20-year sales forecast for manpower planning, and the production manager a 7-year sales forecast for machinery and plant planning, then such forecasts must necessarily be made, however hazardous they may be.

Forecasting techniques are applied to historical data: the more recent the data, the more accurate the forecasts are likely to be. In arriving at a sales-forecasting equation, sales become the dependent variable, the independent variables consisting of those of the demand function (D7). If the company pursues a rational pricing policy (R6), the marginal cost may take the place of price in the equation for each individual product or brand; personal income, the price level of competitors' brands, and expenditure on advertising by the company and competitors are necessary independent variables, for each of which, forecasts must be made year by year.

A simple method to apply is that of trend ratios (T16), which can be made to give greater weight to the more recent historical data compared with earlier data. For consumer goods, especially those of the luxury class, the most important of the independent variables is probably personal income. Sales proceeds of the company's products can be linked statistically to personal income, and forecasts made without the need for an equation. Such sales forecasts can then be adjusted for the forecasts of the other independent variables, and also for expected disturbing factors. Table L14.1 demonstrates the process.

The next step is to predict the future trend ratio of personal income for each year in the future, as far as required. The last two years show increases of more than 10% a year. Is this likely to continue? Or is it not more likely that 8%, the average over five years, will be the appropriate figure? During the preceding five years − from 1960-65 − the average annual increase was 7%, so that the rate of increase itself was tending to rise. A reasonable rate for the years 1972-76 seems to be 9%; for all subsequent years, 10%. A disturbing factor in 1972 and 1973 was the prices and incomes freeze which operated from October 1972, and continued in 1973. The freeze undoubtedly had the effect of reducing the trend ratio in 1972, and also probably in 1973. This is mentioned to indicate the kind of disturbing factors that must be considered as possibilities.

Table L14.1   Trend ratios applied to sales forecasting.

Year	Personal income of the UK		Sales proceeds from product		(4) ÷ (2)
	£ million	Trend ratio	£'000	Trend ratio	
	(1)	(2)	(3)	(4)	
1966	32.1	–	436	–	–
1967	33.8	1.053	467	1.070	1.016
1968	36.3	1.074	510	1.092	1.017
1969	38.8	1.069	556	1.090	1.020
1970	42.8	1.103	624	1.122	1.017
1971	47.1	1.101	702	1.125	1.022
Compounded annual average		1.080		1.100	1.019

(1) Source: National Income, blue book for 1972, HMSO.

Of far greater importance, however, are the causes of the great increase in personal income during the years from 1962 to 1971, and whether they are likely to continue to operate. Will the Government succeed in its efforts to contain inflation and so reduce the rate of increase? Or will it defeat its anti-inflation objectives by its efforts to reduce unemployment? Nobody can answer such questions, but they show the need for caution in forecasting so important an environmental factor as personal income. An annual increase of 6 or 7%, ie a trend ratio of 1.06 or 1.07 may ultimately prove to be more realistic than one of 9 or 10%.

Once the personal-income forecast trend ratios have been made, the ratio of ratios in column five of Table L14.1 can be brought into play to give the forecast trend ratios of sales proceeds from the product. It may be suggested that the annual average of 1.019 be adopted for the purpose.

The forecast trend ratios of personal income are multiplied by

1.019 to obtain them. The forecast sales proceeds can then be built up, year by year, by applying their forecast trend ratios to the latest available actual figure of sales proceeds.

It goes without saying that the forecast figures of personal income can be used in a forecasting regression equation of demand (R7).

The Society for Long Range Planning is devoted to the scientific development of long-range planning techniques and to promulgating its benefits to business management. The Society promotes meetings and conferences and publishes the quarterly journal *Long Range Planning.*

m

M1   Make or buy?

A question of policy relating to parts and components based on an investment appraisal and a forecast of requirements. If, according to the investment appraisal, the annual cost of making a component (C28) is

$$T_1 = aQ + F \tag{1.1}$$

where a is the marginal cost, Q the annual quantity that may be required, F the additional fixed cost and the annual cost of purchasing the same quantity is

$$T_2 = pQ \tag{1.2}$$

where p is the price; then, where T_1 exceeds T_2, it would pay to buy, and where T_2 exceeds T_1 it would pay to make.

Let Q_b be the break-even quantity where $T_1 = T_2$, then

$$aQ_b + F - pQ_b = 0$$

Therefore $\qquad Q_b = \dfrac{F}{p - a} \tag{1.3}$

which is the break-even formula. Where Q exceeds Q_b, it would pay to make, and where it is less than Q_b, it would pay to buy. A diagramatic representation is given in Figure B14.1, in which the graph of (1.1) above is marked T, that of (1.2) is marked S, and $Q_b = Q_v$ on the horizontal axis. The shaded areas labelled *loss* and *profit* should be ignored for make-and-buy policy purposes. See T13.

Applications.   Purchasing policy, investment programming.

M2   Man-hours function

The same as the labour-time function (L1).

M3   Marginal analysis

Analysis by the calculus of finite differences (F3) or the infinitesimal calculus as in M6, as distinct from aggregative analysis (A7).

M4   Marginal cost

The net additional cost, to the company, of increasing the rate of output by one unit; since it remains constant for all rates of output within the normal capacity of the plant (P17), it is equal to the annual variable cost (V5) divided by the annual output up to the normal capacity. Where the normal capacity is exceeded, the marginal cost begins to rise until it reaches a point at which it is equal to the unit cost. This is the economic capacity of the plant, where any further increase in the rate of output results in a reduction of profit. However, this is complicated by pricing policy (P17, R6).

Table M4.1   A marginal-cost schedule.

Annual output		Annual cost		Marginal cost	Unit cost
Total '000	Difference '000	Total £'000	Difference £'000	(4) ÷ (2)	(3) ÷ (1)
(1)	(2)	(3)	(4)	(5)	(6)
10	–	60.00	–	–	6.00
20	10	90.00	30.00	3.00	4.50
30	10	120.00	30.00	3.00	4.00
31	1	123.10	3.10	3.10	3.97
32	1	126.41	3.31	3.31	3.95
33	1	129.93	3.52	3.52	3.94
34	1	133.66	3.73	3.73	3.93
35	1	137.60	3.94	3.94	3.93
36	1	141.75	4.15	4.15	3.94

A hypothetical marginal-cost schedule for rates of output rising from below the normal capacity to the economic capacity is given in Table M4.1. It will be seen that the normal capacity is 30 000 units a year with a marginal cost up to that point of £3.00, and that the economic capacity is slightly less than 35 000 units a year, where the marginal cost has risen to an average level equal to the unit cost. It should be noted that in the table the marginal cost is the average for each range of the annual outputs, that of £3.10, for instance, is the average for the range of output rates of 30 000 units to 31 000 units.

The marginal cost of the 30 001st unit would be rather less than the average but slightly more than £3.00, and that of the 30 999th unit would be rather more than £3.10 and rather less than £3.31. On the other hand, each unit-cost figure relates directly to the corresponding rate of output given in column 1. However, since it varies little at the critical range of outputs from 34 000 to 36 000, a comparison with the marginal-cost figures at this range is valid.

Something needs to be said about a common misconception regarding the marginal cost, a misconception which is often found in places where one would least expect it. It is a dichotomy of the marginal-cost concept between what is called the *short-term marginal cost* and what is called the *long-term marginal cost*. The marginal cost discussed above and calculated by finite differences (F3) in Table M4.1 would be classified as the short-term marginal cost.

The long-term marginal cost is supposed to take account of the cost of increases in the capacity of the plant. As a simple calculation would show, the annual fixed cost of the product exemplified in Table M4.1 is £30 000. When the rate of output is equal to the economic capacity of the plant, the firm, let us suppose, increases the annual capacity by 5000 units at an additional annual fixed cost of £6000, with no change in the marginal cost. It then looks as though the annual cost curve rises suddenly, by £6000, up the ordinate of the 35 000th unit of output. But this is quite illusory; immediately the extension is brought into use, the whole cost curve rises bodily throughout its whole length to a new position – equal to £36 000 on the vertical axis – and slopes upwards to the right, parallel to the original cost curve. The additional annual fixed cost of £6000 remains *fixed:* if the rate of output were to fall to 20 000 units a year, the total annual cost would be £96 000, not £90 000, as it was before the plant was extended; the same applies to all other rates of output. How can the additional fixed cost of £6000 a year be conceived as forming part of the marginal cost without destroying the whole concept? It simply cannot be done; the marginal cost is the unit variable cost: the 'long-term' argument of the dichotomists is a contradiction in terms.

Another difficulty sometimes arises when a plant is extended. The new plant may produce at a lower marginal cost than the original, but still extant plant. Here the cost curve appears to be less steep for the higher rates of output. But, in fact, since it would be more economic to use the new plant to its normal capacity and to allow the original plant to provide the spare capacity, the opposite is probably true: the cost curve becomes less steep for the lower rates of output, and rises more steeply at the ordinate representing the normal capacity of the new plant. Provided the rate of output is great enough to necessitate using the original plant, the true marginal cost for most purposes, including price fixing, is that of producing by the original plant.

Where two or more products or brands are involved, there is another problem. It is possible to calculate the marginal cost for each product by statistical analysis rather than cost-accounting synthesis. The difficulty with synthesis is that it implies a definition of product variable cost by connotation: direct wages and materials, fuel, power and part of repairs. Generally, direct wages and materials costs are the only variable costs that can be definitely ascribed to individual products. In any event, there is often a variable element in materials handling, warehousing and distribution costs, but it can rarely be ascertained except, possibly, by statistical analysis.

As stated in C28, a regression analysis (R7) of annual costs based on the model of (C28.2) could rarely be used where the individual products exceed four or five. There would not be enough data. An alternative approach is to compile an index number of output weighted by reference to the connoted marginal cost (as far as it can be ascertained with certainty), to provide the series of the independent variable. Laspeyres's index (L4) could be used for the purpose. Total cost in unit time, preferably excluding such hard-core fixed costs as rent of buildings, local rates and top management fees and salaries, would provide the dependent variable. Then Equation (C28.2) would form the model with Q as the index of production. When the regression equation has been determined, the next step is to calculate the extent to which each product contributes on average to a rise of one point in the index.

Suppose a firm produces four products for sale A, B, C and D, and finds that the regression coefficient of the output index is 300. If the dependent variable, total cost as defined, is expressed in terms of £s, then the coefficient means that the total cost in unit time increases by £300 for every rise of one point in the output index. Table M4.2 demonstrates the procedure for estimating the total marginal cost of each product.

Table M4.2   Estimating the marginal cost.

Product	Weight W	Output Q	WQ $=V$	$\dfrac{\Sigma V}{100W}$	Estimated marginal cost £300 ÷ (4)	Check $\dfrac{(2) \times (5)}{100}$
	(1)	(2)	(3)	(4)	(5)	(6)
A	1.0	1 000	1 000	165.00	1 818	18.18
B	2.5	3 000	7 500	66.00	4 545	136.35
C	2.0	2 000	4 000	82.50	3 636	72.72
D	4.0	1 000	4 000	41.25	7 273	72.73
Total			16 500			299 98

Both W and Q apply to the base period of the output index; W represents the 'connoted' marginal cost of each product, and Q the quantity made. The sum of WQ, shown as ΣV, is equivalent to the base-period weight of Laspeyres's index. To make the index, it is multiplied by the quantity relative, Q_n/Q_0, for each product totalled and divided by the sum of W_0Q_0, equal to 16 500 in the example. In the course of index number making, the figure of 16 500 is equated to 100. The extent of the increase of ΣV for a rise of one point in the output is 0.01 of 16 500 = 165, which, divided by W, gives the increase in the output of any one product necessary to raise the output index by one point. The increase for each of the four products is shown in column 4. Dividing this into the figure of £300, we have the estimated marginal cost of each product. Column 6 is added mainly as a check on the method and the arithmetic. It should total 300, the slight difference being due to rounding.

It will be seen by comparing columns 5 and 1, that the method in effect allocates the 'unconnoted' marginal costs *pro rata* to the 'connoted' marginal costs, to satisfy the denoted definition of *marginal cost*. It is somewhat arbitrary, but it is the best that can be done with the material available. It is important to see that the 'connoted' definitions in respect of the several products are comparable; that they cover the same items.

Applications.   Variable costing, product costing, rational price fixing.

M5   Marginal efficiency of capital
The internal rate of return (I14).

M6   Marginal revenue
The change in gross revenue due to a change of one unit in the rate of sales resulting from a change in price. Its formula is

$$M = p(1 - \frac{1}{e})$$ (6.1)

where M is the marginal revenue, p the price, and e the price elasticity of demand. The price elasticity of demand is an econometric concept of considerable practical significance. Its formula is

$$e = \frac{dQ}{dp} \cdot \frac{p}{Q}$$ (6.2)

where e is constant for a given price range. The demand function is that of Equation (D.1), from which is derived:
$$p = kQ^{1/e}$$ (6.3)
The corresponding cost equation (cost of sales) is
$$T = aQ + F$$ (6.3)
where T is the total cost of sales, and a the marginal cost (M4).

Since the marginal cost is the net additional cost of an increase in the rate of output of one unit, the marginal *net* revenue, N, may be written:

$$N = p(1 - \frac{1}{e}) - a$$ (6.5)

Where the marginal cost exceeds the marginal gross revenue, the sign of N is negative, and where it falls short of it, it is positive. It would seem that a company's aim should be to maximize the marginal net revenue, but oddly enough, this is not so. To maximize the total annual net revenue, the aim should be to maintain the marginal net revenue at zero, ie to equate the marginal revenue to the marginal cost. Graphically, the former is the measure at a given point of the slope of the aggregative sales curve, the latter, the measure of the slope of the aggregative cost curve. Figure B14.2 shows that net revenue is at its maximum where the two curves are parallel, ie where $M = a$. The slope of any curve is equal to the differential coefficient of the factor measured on the y-axis with respect to the factor measured on the *x*-axis. From (M6.3)

$$S = pq = kQ^{1-1}/e \qquad (6.6)$$

where *S* is the gross revenue in unit time from the product. We therefore have

$$M = \frac{\cdot dS}{dQ} = (1 - 1/e)kQ^{1}/e \qquad (6.7a)$$

Since $kQ^{1}/e = p$ — see (6.3) — we can rewrite (6.7):

$$M = p(1 - 1/e) \qquad (6.7b)$$

In the marginal net-revenue equation, (6.5), of the three factors of the righthand side, *a* and *e* are constants, and *p* an independent variable. Under conditions of imperfect competition (I1), and Government price freezes apart, the company has control of *p*, the selling price of the product. The price that maximizes the total net revenue from the product, written p_m, is called the optimum price. See R6. As shown above, it exists where $M = a$, ie

$$a = p_m \left(1 - \frac{1}{e}\right)$$

$$\text{Therefore} \qquad p_m = a \frac{e}{e - 1} \qquad (6.8)$$

Thus the concept of marginal revenue is of practical importance in the financial policy of any commercial undertaking.

Applications.   Marketing, rational price fixing.

M7   Market research

Collecting, collating and analysing all the relevant facts available in a given market, or group of markets, for a company's product or service, whether existing or projected. Since the early history of market research, there has been a growing tendency to develop its techniques along scientific lines in which statistical collection and analysis play an important part. Statistical sampling (S3, 4, 5 and 6) has become almost as deeply involved in market research as it is in quality control (Q3, S30).

Modern market research has its philosophy, as well as its techniques. It is called *free form,* or sometimes *free-form management.* Free form was developed by an American industrial research organization called Equity Research Associates (ERA), and was first adopted by Litton Industries, a growth conglomerate, which recently fell on hard times. ERA propounded free form as a new business philosophy whose primary faith lay in promoting the general welfare rather than in trying to make money. Perhaps the money would flow into the company's coffers in spite of it, or in consequence of it. Market research workers would go out into the world, not so much to seek markets for existing products as to seek gaps in the public welfare, and return home to find the best economic means of filling them. It would sound cynical to add — *at a profit* — but there is a limit to philanthropy. If free form has proved to be successful, its success may lie in the fact that one's own welfare depends very largely on the welfare of one's customers. Promoting one's customers' welfare amounts to promoting one's own.

Market research is also concerned with determining the effect on the demand for the company's products, of changes in selling price (R6), and of other factors such as personal income, advertising and competitors' behaviour. In short, it is one of the tasks of market research to derive a regression equation (R9) based on an appropriate demand function (D7), for each product brand or service sold by the company.

It is also concerned with product design, methods of distribution, quality control, and many other aspects of business management. Some of its work is routine, but probably most is *ad hoc*. Members of the research team need to have creative, as well as analytical minds, and to be resourceful and ingenious.

M8   Mathematical model
See model building (M16).

M9   Matrix
In this context, a statistical table with two or more dimensions, eg the input-output matrix. A way of arranging numerical information in a systematic way to facilitate retrieval and to save repetition.

Matrix algebra. A system of algebra in which basic data and symbols for unknown quantities are set out in matrices; said to be of advantage in solving simultaneous equations in which the number of variables is large, but has also been called a mathematicians' plaything.

Matrix management. A form of management in which the

technical management is decentralized and responsible for local routine operation, but reports to the head technical manager in the field at headquarters.

M10   Maximum likelihood
See likelihood (L9).

M11   Mean
Average (A14 and 16, G2, H1, M14 and 15).

M12   Mean deviation
Average deviation (A7); also deviation from the mean (D19).

M13   Mean-square deviation
The square of the deviation of a figure of a series from the mean of the series; used extensively in mathematical statistics, including correlation, least-squares regression analysis and calculating the standard deviation and the standard error.

M14   Median
The middle figure of a series arranged in order of magnitude; it is a type of average and a measure of central tendency. See A16, C10.

M15   Mode
In a frequency distribution, the item that occurs the most frequently; it is a type of average and a measure of central tendency. See A16, C10.

M16   Model building
Setting out on paper the system of an organization or a whole economy; drawing up the order of procedures or processes in a workshop or office or on a building site; setting out the chain of cause and effect or the relationship between factors. Of the three main types of model — (1) the verbal; (2) the diagrammatic; (3) the mathematical — the statistician is concerned mainly with the last in respect of the relationships between factors. A mathematical model is essential in multiple correlation and in regression analysis. The techniques and theoretical considerations that can be applied to determining the appropriate mathematical form of model are

discussed under graphing (G5), cross-classification (C36) and finite differences (F3).

M17   Modulus
A number expressed in terms of its value without regard to its sign. The average deviation (A17) is derived by using the moduli of mean deviations. In some contexts it means a mathematical constant.

M18   Moments
A term that is tending to fall into disuse as superfluous; still seen occasionally in current literature in such phrases as the *method of moments* and *moments about the mean.* There are three moments:
1   The average deviation (A17).
2   The standard deviation (S24);
3   A measure of skewness which is sometimes called the third moment (S20). See also F9.

M19   Money, value of
Since the be-all and end-all of all economic activity is the satisfaction of human wants, the value of money is determined by the quantity of consumer goods a given sum will purchase in one period compared with another, the base period. Its measure is the inverse of the retail-price index (C30).

M20   Monte Carlo method
A simulation technique (S17) originating in operational research (O4) designed for solving queueing problems (Q7), ie for finding the organizational method that would probably minimize the waiting time of men and things in such situations as ships waiting for a vacant berth in a port or an aeroplane waiting for permission to land. The technique employs mathematics and probability theory applied to a number of proposed alternative organizational methods.
   Applications.   Investment programming, factory reorganization.

M21   Moving averages
A method of smoothing a time series and for eliminating seasonal variations (S11). It is not a method of regression analysis (R7), but can be used effectively in tandem with finite differences (F3) in empirical model building (M16) and for solving linear-regression problems (L11).

A moving total is first obtained, and from this, the moving average is calculated. For a monthly series, the moving total and monthly average sales are worked out over each successive month; for instance, the totals for January, February and March, then for February, March and April, and so on, are determined and averaged. This is called a three-month moving average. An example of a three-month moving total and moving average is given in Table M21.1. The smoothing effect is seen in a comparison of the range of the moving averages, 373 − 273 = 100, with that of the original series, 510 − 220 = 290. The twelve-month moving average is based on twelve consecutive months. It therefore has the effect of removing seasonal variations. However, like other statistical devices, the method needs to be used with care and discretion. A twelve-month moving average certainly removes seasonal variations: it may also remove, or at least hide, changes for which one might be looking. For instance, there may be a non-seasonal fall in sales which is lost in the moving figures. The moving figures are, by definition, less sensitive to seasonal and other influences. It is theoretically possible for a series of moving figures to be all equal, or to consist of an arithmetic or geometric progression. Consider the figures in Table M21.2, for instance: the moving totals, and therefore the moving averages, are geometric progressions with a common ratio of 1.20.

Table M21.1 Three-month moving total and moving average of sales. £'000

		Monthly figures	Moving total	Moving average
1972	August	510		
	September	250		
	October	360	1 120	373
	November	240	850	283
	December	220	820	273
1973	January	430	890	297
	February	370	1 020	340

Both sets of moving figures reveal nothing more than that the original series is tending to increase by a constant percentage of 20% a month, though one would scarcely think so from the original series itself. The question is whether the monthly figures themselves are not more revealing of the information required for the control function than are the moving figures, which are really trend norms.

It should be kept in mind when drawing conclusions from moving figures that, if the first original figure of the previous total is on the high side, and the last figure of the current total is on the low side, or vice versa, a comparison of the moving figures for the current month with the previous month might be misleading: the moving total and average for the current period would show a greater decrease or a smaller increase compared with the previous period than the original figures show.

Table M21.2 Theoretical three-month moving figures comprising geometric progressions.

		Monthly figures	Moving total	Moving average
1972	August	400		
	September	290		
	October	310	1 000	333
	November	600	1 200	400
	December	530	1 440	480
1973	January	598	1 728	576
	February	946	2 074	691

In its use in tandem with finite successive differences, the method has the effect of reducing variations in the estimates of the coefficient of the independent variable and thus renders them more amenable to the drawing of inferences about the shape and slope of the line of best fit (B5). The two methods can be applied in tandem to the solution of simple-regression problems where the independent variable is either time or any other factor. For a discussion of the former, see time-series analysis T8.

Applications. Analysis of time series; useful in model building.

M22   Multi-asset project

A proposed capital work in which it is planned to provide two or more fixed assets with different book lives, so that separate provision for renewal or amortization of each asset has to be made. See replacement of fixed assets (R11) and sinking fund (S18); also A12, I14, N4.

M23 Multiple-regression analysis

Regression analysis (R7) involving two or more independent variables. Of the methods of multiple-regression analysis, least squares (L7) and cross-classification (C36) are among the best.

M24 Multiplication law

A law of chance in which two or more probabilities are multiplied together to give a total probability that each of two or more events will occur. For instance, if a race-course punter bets on an accumulator of three horses, then, in the betting market's estimation, the probability that the three horses will win is the product of the odds. If they are 2 to 1 against, 5 to 4 against and 9 to 4 against, the probability that all three horses will win is

$$\frac{1}{3} \times \frac{4}{9} \times \frac{4}{13} = \frac{16}{351} = \frac{1}{22} \text{ or 21 to 1 against}$$

The multiplication law has its most valuable use in determining the probability of finding two or more unrelated characteristics together in one person or thing. In quality control, for instance, it may be considered highly undesirable that two kinds of defect should be found together. It is known that of a batch of 10 000 articles, 1% have one defect, and 2% the other defect. What are the chances that the two defects will be found together — 1% of 10 000 is 100; the probability is that 2% of this 100 will also have the second defect, ie it will occur in two of the batch of 10 000, equal to 0.02%. Applying the multiplication law to the two probabilities, we have

$$\frac{1}{100} \times \frac{2}{100} = \frac{2}{10\ 000} = 0.02\%$$

There are two laws of chance, the other being the addition law (A5).

M25 Mutual regression

A notional formula of the method of least squares (L7) in which the burden of error would be distributed equally among the several variables. In the regression of Y on X, for instance, Y carries the burden of error, ie Y and not X is made subject to all the variations due to factors other than X in the relationship between the two factors. Mutual regression would take a middle course between the regressions of Y on X and X on Y.

In the early fifties, the problem of finding a mutual-regression formula was examined by the statistics section of the United Nations Organization but, so far as is known, no solution was found. The matter is discussed in some depth under least squares (L7): no algebraic formula is offered, but a more flexible solution is described for regression problems where intercorrelation exists.

n

N1   National income

The gross national product less capital consumption; sometimes called the net national income. Capital consumption is much the same as depreciation in the commercial accountancy sense of the term; it is officially defined as 'the amount of fixed capital used up in the process of production'. It is the value of that part of the gross domestic fixed capital formation required to maintain the national stock of fixed capital assets.

    Sources.   For annual figures *National Income and Expenditure,* annually, HMSO; for quarterly figures *Monthly Digest of Statistics,* and the January, April, July and October issues of *Economic Trends,* both HMSO. Methods, definitions and sources are described in *National Accounts Statistics: Sources and Methods,* which is No. 13 of *Studies in Official Statistics,* HMSO, 1968.

N2   Negative correlation

Two statistical series are said to be negatively correlated when as one rises, the other tends to fall. Unit costs and the rate of output are negatively correlated, so, also, are selling price and demand; output and absenteeism; productivity and labour turnover; sales and competitors' expenditure on advertising. The concept applies only to pairs of factors, ie to simple correlation. It does not apply to multiple correlation. For formulae and numerical examples, see correlation (C25).

N3   Net output

Much the same as value added (V1); it is equal to the sales proceeds less the cost of purchased materials, parts, components and fuel. The term originated in the censuses of production. Net output is additive, firm by firm and industry by industry, whereas sales proceeds cannot be added without incurring duplication. The use of net output as a factor in making an index number of productivity is examined under productivity (P33).

N4   Net present value

One of the three principal techniques and yield criteria of discounted cash flow; it is equal to the present value of the net yield of a fixed asset over its life less the outlay on the asset. In the discounting process an appropriate current rate of interest is used. The lower the rate of interest used for the purpose, the greater is the calculated present value of the yield. The present value of £1000 a year for 20 years at 15% is £6259; at 10% it is £8514; and at 5%, £12 462. The net yield of the asset should be gross of depreciation, as the NPV technique embodies sinking-fund (S18) depreciation provision and also charges interest on capital — both at the rate of interest used in the discounting process.

Suppose the outlay on the asset is £5000 and the rate of interest 10%, then the NPV would be £8514 − £5000 = £3514. This is net of interest on capital and depreciation:

	£
Interest on £5000 at 10%	500
Depreciation sinking fund:	
20 years at 10%: £5000 at 0.01746 per £	87.3
Annual capital charge	£587.3
£1000 a year less £587.3 a year	£412.7

The present value of £412.7 a year for 20 years discounted at 10 per cent is

$$£412.7 \text{ at } £8.514 \text{ per } £ = £3514$$

which is the net present value of the asset.

As a yield criterion rather than as a technique, NPV has certain properties which are discussed under yield criteria (Y3).

Applications.   Investment appraisal where projects provide for a single fixed asset.

N5   Net regression coefficient

The coefficient of an independent variable in an analysis involving at least one other independent variable; any coefficient in a multiple-regression equation (R7, 8 and 9). It is described as *net* because it measures the effect on the dependent variable of changes in an independent variable net of the effect of changes in other independent variables. In recent years the phrase seems to have been falling into disuse as redundant to requirements.

N6   Net-revenue schedule

A statistical schedule which sets out a range of prices of a brand of product, and the corresponding annual sales in terms of quantity and money, the annual total or variable costs and the profit. A demand schedule is contained in Table D8.1, and it is proposed here to extend it to make a net-revenue schedule by adding a cost schedule (C32), in which the marginal cost is £6, and the annual fixed cost £3000. Table N6.1 is the result.

Table N6.1   Hypothetical annual net-revenue schedule.

	Annual sales		Annual cost		Annual
Price	Quantity	Proceeds	Variable (2) x £6	Total (4) + £3000	profit (3) − (5)
£	no	£	£	£	
(1)	(2)	(3)	(4)	(5)	(6)
10	1 000	10 000	6 000	9 000	1 000
11	826	9 086	4 956	7 956	1 130
12	694	8 328	4 164	7 164	1 164
13	592	7 696	3 552	6 552	1 144
14	510	7 140	3 060	6 060	1 080
15	444	6 660	2 664	5 664	996

It will be seen that the optimum price is £12, which agrees with that arrived at by the optimum-price formula given in M6:

$$p_m = 6\frac{2}{2-1} = £12$$

It will be seen in D8 that, for this example, the elasticity of demand is 2.

Applications. Presentation   purposes,   financial   planning, advertising policy.

N7　Non causa

False cause; a term used in logic for the fallacy of assuming an unproved cause. See causal relationship (C7) and correlation (C25); also N8.

N8　Non sequitur

Invalid inference; a term used in logic, and meaning 'does not follow', for a number of fallacies including false analogy and false preciseness. False preciseness is not uncommon in statistical and economic writing. A typical example is:

> The Board's original off-form cocoa price . . . was set at a period when the world price of cocoa was low, and was therefore, between 15 and 20% lower than that originally offered.

The whole sentence is somewhat woolly, but its outstanding fault lies in the precise inference drawn from imprecise premises. It should be noted that the truth of the inference is not in question; the point is, it does not follow from the stated premises.

N9　Normal curve of error

A symmetrical frequency-distribution curve (F9); variously called the *normal probability curve,* the *normal frequency-distribution curve* and, in its context, simply the *normal curve.* Although the binomial distribution curve (B11) and the Poisson distribution curve (P19) are of importance in statistical work, such as quality control, neither compares in importance with the normal curve in statistical theory and the practical applications of the law of errors, as the normal curve means what its name implies: most frequency distributions found in practice conform to it near enough.

Its mathematical formula is

$$y = \frac{1}{\sigma\sqrt{2}}\, e^{\dfrac{-(x-\bar{x})^2}{2\sigma^2}} \tag{9.1}$$

where y is the height of the curve for any given value of x measured on the x-axis, $\bar{x}$ is the average of x, π is the ratio of the circumference of a circle to its diameter, ie 3.1416, and e is the base of Naperian logarithms, ie 2.7183. It will be appreciated that, as π and e are constants in the equation, all we need to know about a normal frequency distribution is the average x and the standard deviation, σ to enable us to estimate the value of y for any value of x.

The equation is stated in a form that makes the total probability (measured by the area beneath the curve) equal to unity. The area beneath the normal curve between two verticals, each one standard deviation from the mean on either side, is equal to 0.6826 of the whole area, which means that any observed value of x has a 0.6826 probability that it will fall within one standard deviation of the mean, and that, of 10 000 observations with a mean of 100 and a standard deviation of 30, it is probable that 6826 of them would fall within the range 100 ± 30, ie between 70 and 130. Between two verticals, each two standard deviations from the mean, the area under the normal curve is 0.9544 of the total area, and between two verticals, each three standard deviations from the mean, the area is 0.9974 of the total. Table N9.1 sets out the percentage probabilities for a range of values of the standard deviation, from 0.1 to three standard deviations.

Table N9.1 Normal distribution probabilities relative to the mean ±nσ.

Number of standard deviations (n)	Probability (%)	Number of standard deviations (x)	Probability (%)
0.1	7.96	1.6	89.04
0.2	15.86	1.7	91.08
0.3	23.58	1.8	92.82
0.4	31.08	1.9	94.26
0.5	38.30	2.0	95.44
0.6	45.14	2.1	96.42
0.7	51.60	2.2	97.22
0.8	57.62	2.3	97.86
0.9	63.18	2.4	98.36
1.0	68.26	2.5	98.76
1.1	72.86	2.6	99.06
1.2	76.98	2.7	99.30
1.3	80.64	2.8	99.48
1.4	83.84	2.9	99.62
1.5	86.64	3.0	99.74

It will be appreciated that the area enclosed by the vertical raised from the mean of the series and the vertical from any other point on the x-axis is half the figure shown in Table N9.1 in the column headed 'Probability'. From this, it is a simple matter to build up a normal frequency distribution for any series. Half the total frequency of 100% is almost entirely accounted for by three standard deviations from the mean in each direction. If the mean of a sample is 60, and the standard deviation is 4, then nearly 50% of the total frequency lies between $60 - 3 \times 4, = 48$, and 60, and the other 50% between 60 and 72. Similarly, for one standard deviation, 34% lies between $60 - 4, = 56$, and 60, and also between 60 and 64. The frequency between one and two standard deviations, is $48 - 34 = 14$.

Table N9.2 Building up a normal frequency distribution: mean = 60, and σ = 4.

Sampling observations	Number of standard deviations from (−) or to (+) sample average	Area of graph covered (%)	Frequency (successive differences)
48-50	−3.0	50	1
-52	−2.5	49	1
-54	−2.0	48	5
-56	−1.5	43	9
-58	−1.0	34	15
-60	−0.5	19	19
-62	+0.5	19	19
-64	+1.0	34	15
-66	+1.5	43	9
-68	+2.0	48	5
-70	+2.5	49	1
-72	+3.0	50	1

The complete distribution is given in Table N9.2. The table is built up on class intervals of 2, ie half a standard deviation. The sample on which it is based may consist of fewer than 10 observations. Provided the sampling average is 60, and the standard deviation, 4, the figures in Table N9.2 form the frequency distribution which one would expect to find in the universe. But see Bessel's correction (B4) and sampling (S3, 4 and 5).

N10   Notation

Statistical notation has become fairly standardized over the years. Processing signs (+ x, etc.) and signs of comparison (= $>$, etc.) are the same as those used in everyday arithmetic and mathematics except that Σ is used for summation, instead of S which serves in statistics as a notation symbol (see below).

Symbols

$(\Sigma x)^2$	square of the sum of x
$\Sigma(x^2)$	sum of the squares of x
$\Sigma(xy)$	sum of the products of the pairs of x and y
$\Sigma x . \Sigma y$	product of the sum of x and the sum of y
χ^2	test of significance (C12)
σ	standard deviation (D19)
S	standard error of estimate (S25)
β	beta coefficient (B6)
r	simple-correlation coefficient (C25)
R	multiple-correlation coefficient (C25)
ΔX	increase or decrease in X
δ	average deviation (D19)
μ	average of universe
$\hat{Y}$	estimate of Y from regression equation
X	average of X
M_x	(mean of X) same as $\overline{X}$
var (X)	variance of X (A10, V7)
n	number of observations in a sample
m	number of variables in a regression equation (R9)
$n-m$	number of degrees of freedom (D6)
z	residual error, ie actual less estimate (S25)

Subscripts in regression analysis by least squares (L7)

$R_{1.234}$	multiple-correlation coefficient where X_1 is the dependent variable, and X_2, X_3, X_4 are the independent variables
$a_{12.34}$	regression coefficient of X_2 for regression of X_1 on X_2 X_3, X_4
$a_{2.134}$	coefficient of X_2 for regression of X_2 on the rest, with X_1 serving as the dependent
$a_{23.14}$	coefficient of X_3 for regression of X_2 on the rest, with X_1 serving as the dependent
$a_{32.14}$	coefficient of X_2 for regression of X_3 on the rest, with X_1 serving as the dependent; and so on

$k_{1.234}$ unattached constant in regression equation for regression of X_1 on the rest

$k_{2.134}$ unattached constant in regression equation for regression of X_2 on the rest, with X_1 serving as the dependent

For simple-regression analysis, Y is generally used to indicate the variable serving as the dependent, and X the independent. Then:

a_{yx} coefficient of X for regression of Y on X
a_{xy} coefficient of X for regression of X on Y

Similarly with the unattached constant.

N11   Null hypothesis

A hypothesis being, or to be, tested. The test may show a hypothesis to be false for a given probability, but not true. If the specified average weight of metal in a series of iron castings of a given design is 10 lb, a sample consisting of, say, 65 items drawn from a batch may be used to test the null hypothesis that the actual average weight is indeed 10 lb. If the average weight of the sample turns out to be 9.5 lb with a standard deviation of 1 lb, the sampling error, s (S4) is

$$s = \frac{1}{\sqrt{(65-1)}} = \frac{1}{8} = 0.125$$

Twice the sampling error gives a 95% probability that the difference between the specified average and the sample average did not happen by chance. These limits in the example are

$$10 \pm 0.25 \text{ lb, i.e. } 9.75 \text{ and } 10.25 \text{ lb}$$

But the sample mean of 9.5 lb lies outside these limits, so that the null hypothesis that the specified mean of 10 lb is consistent with the sample mean of 9.5 lb has to be rejected.

The test can be applied in another way. It is to divide the standard error into the difference between the specified and sample means, which gives the measure of the difference in terms of standard errors:

$$\frac{10.0 - 9.5}{0.125} = \frac{0.5}{0.125} = 4$$

Since the ratio exceeds 2, the null hypothesis must be rejected, the conclusion being that the iron castings comprising the batch from which the sample was drawn were consistently and significantly under weight.

Applications. Testing weight, length and volume of the goods purchased or goods to be sold against specification or other standards of comparison under a system of quality control.

O

O1   Objectives
The aims of any commercial or non-commercial undertaking. It is a useful working hypothesis to accept the principle that the primary objective of a commercial undertaking is to make a profit and to maximize it. To achieve this, a number of subservient or secondary objectives have to be achieved, and there is a growing belief that to recognize them, classify them and seek the best means of achieving them form an important function of management; hence such well-known management philosophies as management by objectives and management by perception. The means of achieving secondary objectives form the body of management techniques, in most of which statistics play at least an important role, and sometimes constitute the complete technique. It would be possible to build a pyramid of commercial objectives with the primary objective at its apex, second-order objectives immediately below, supported by third-order and even fourth-order objectives. On the financial-policy side, for instance, we would have rational pricing (R6) as one of the second-order objectives, with third-order objectives consisting of calculating the elasticities of demand (D7) and the marginal cost (M4), supported by such fourth-order objectives as collecting and collating the relevant basic statistics, and so on.

Applications.   Reorganization.

O2   Obsolescence
Approaching obsoleteness of fixed assets due to improved technology and changing taste and fashion. See premature displacement of assets (P22).

O3   Ogive
The graph of a cumulative frequency distribution. See C38 and F9.

O4   Operational research
Mathematics and economic theory applied to the field of business management and control, the object being to provide data for policy and executive decision making. Includes linear programming (L10), break-even analysis (B14) and profit planning (P26). Some of the mathematical techniques were in management use in the thirties, long before they were blessed with a name.

O5   Opportunity cost
The cost of a lost opportunity, sometimes called *opportunity loss;* generally applied in management and accounting literature to the yield of marketable securities that a company sells to provide capital finance for the acquisition of fixed assets. The securities and their yield may be notional rather than real. The opportunity cost is usually taken care of in the interest and depreciation charges.

Applications.   Investment appraisal, efficiency auditing.

O6   Optimum value
That value which has the effect of maximizing another value, eg the optimum price (R6) and the EOQ (E6), both of which have the effect of maximizing profit.

O7   Ordinate
The vertical scale or axis of a graph. See y-axis (Y1).

O8   Overhead
Of the total annual cost, that part which does not vary with the rate of output, but exists in time. Variously called *fixed costs* (F8), *oncosts, burden, time costs* and *period costs.* The literature is often somewhat vague about the definition. In its *Terminology of Cost Accountancy,* the Institute of Cost and Works Accountants, does not equate overhead to fixed costs, but defines it as 'the aggregate of indirect materials cost, indirect wages (indirect labour cost) and indirect expenses'. It defines indirect costs as costs 'which cannot be allocated but which can be apportioned to, or absorbed by, cost centres or cost units.' Whether the ICWA's definition of overhead is

consistent with such phrases as *variable overhead,* which appear in the literature, is uncertain. It is a term that is best avoided. See also A3 and 15, C26, V5.

O9   Overseas trade

The visible (V12) and invisible (I18) trade of the country. Visible imports are classified by commodity and country of origin in accordance with the *Import List* (I3), and visible exports, in accordance with the *Export List* (E11). Invisibles, which include capital transfers, are classified under a number of main headings.

Sources.   Visible trade of the UK: for monthly and cumulative statistics, *Annual Statement of Trade: Overseas Trade Accounts,* formerly the *Monthly Trade and Navigation Accounts,* all HMSO. Both the *Annual Statements* and the monthly *Accounts* give an appreciable amount of detail, the former rather more than the latter. It is possible in some circumstances to obtain unpublished figures from the Bill of Entry Service, Customs and Excise, King's Beam House, Mark Lane, London EC3.

Invisible trade of the UK: *Financial Statistics,* HMSO, Bank of England *Quarterly Bulletin.*

P

P1   Paasche's index number

One of the two basic index numbers of the Laspeyres-Paasche group (I7). It is similar in form to Laspeyres's index number, but is currently weighted instead of base weighted. The aggregative formula for a price index is

$$P_{n.o} = \frac{\Sigma(p_n q_n)}{\Sigma(p_0 q_n)} \tag{1.1}$$

where $P_{n.\,0}$ is the index for the given year, n to the base year 0, $= 1.00$; p is the price of a given item, q the quantity of the given item, and Σ the sign of summation. Since the value, pq, of any item is nearly always much easier to come by than the quantity, q, there would be little point in converting the aggregative form to an arithmetic one, for which the value weight would consist of the quantity in the given period, valued at prices in the base period. Such a conversion would necessitate using the inverse price relative (P26), ie p_0/p_n, instead of $p_n p_0$. However, an harmonic form is practicable, namely

$$P_{n.o} = \frac{\Sigma(p_n q_n)}{\Sigma\left(p_n q_n \dfrac{p_0}{p_n}\right)} \tag{1.2}$$

again necessitating the use of the inverse price relative.

Paasche's quantity index formulae are:

Aggregative

$$Q_{n.0} = \frac{\Sigma(p_n q_n)}{\Sigma(p_n q_0)} \tag{1.3}$$

Harmonic

$$Q_{n.0} = \frac{\Sigma(p_n q_n)}{\Sigma\left(p_n q_n \frac{q_0}{q_n}\right)} \tag{1.4}$$

Although Paasche is generally considered to be as good as Laspeyres, it is rarely used for making official index numbers, though owing to the method used for calculating quantity index numbers from price index numbers, ie by dividing them into the value index,as described under Index numbers (I7), the quantity index thus made is Paasche's.

Paasche's index is not necessarily a chain-base index: the four formulae of (6.1) to (6.4) above are all fixed-base formulae. Its chain-base form is given under chain-base index numbers (C11). See also F6, S38.

Applications.   Testing Laspeyres's index for bias, and for use as a term in Fisher's ideal index.

P2   Parabola

An archlike curve described on the surface of a cone by the intersection, at a place below its apex, of a plane parallel to the opposite side. Now generally used of any archlike curve, such as that described by the flight through the air of an arrow or bullet, and that described by a quadratic equation (Q2) whose coefficient of X^2 has a negative sign. The graph of a frequency distribution (F9) is not called a parabola.

P3   Parameter

Variously defined (1) as a generic term for derived statistics (D11) of all kinds — simple averages, standard deviations, regression coefficients and the like; (2) as a value required for defining a given frequency distribution, ie the average or the standard deviation. The former is the more widely accepted definition.

P4   Part correlation

The statistical correlation (C25) that exists between the dependent and one of the independent variables in a multiple-correlation problem after the effect on the dependent variable of the other independent variables has been eliminated. See also Tables C25.2 and 3. The coefficient of correlation thus calculated (1.00 in the example) is called the coefficient of part correlation. To carry out the eliminating process, it is first necessary to do a multiple-regression analysis. If the problem is in four variables, X_1, X_2 X_3 and X_4, with X_1 regarded as the dependent variable, then the coefficient of part correlation between any two variables, with the eliminating process carried out by reference to the regression of X_1 on the rest (L7), will tend to differ from that when the elimination is carried out by reference to the regression of X_2 or X_3 or X_4 on the rest. The coefficient of part correlation is represented by the symbol $_{12}r_{34}$ for the correlation between x_1 and X_2 based on the regression of X_1 on the rest, and by $_{21}r_{34}$ for the correlation between X_1 and X_2 based on the regression of X_2 on the rest. See notation (N10).

However, the eliminating process can be avoided by using the following formula:

$$_{12}r_{34} = \sqrt{\left(\frac{a_2{}^2\,\sigma_2{}^2}{a_2{}^2\,\sigma_2{}^2 + \sigma_1{}^2(1-R^2)} \right)} \tag{4.1}$$

where a_2 is the regression coefficient of X_2 (L7, R8), σ_1 and σ_2 are the standard deviations (D13) of X_1 and X_2 respectively, and R is the multiple-correlation coefficient (C25). See P5.

Applications. Testing the statistical importance of an independent variable in a multiple-regression analysis (M23).

P5   Partial correlation

The statistical correlation (C25) that exists between the dependent variable and one of the independent variables in a multiple-correlation problem; calculated by reference to the effect on the multiple-correlation coefficient of omitting that independent variable from the analysis. The coefficient of partial correlation is represented by the symbol $r_{12.34}$ for a problem in four variables, where it applies to the correlation between X_1 and X_2, and by $r_{13.24}$, where it applies between X_1 and X_3. See notation (N10). Formulae for these two, given by way of examples, are as follows

$$r_{12.34} = \sqrt{\left(1 - \frac{(1 - R^2_{1.234})}{(1 - R^2_{1.34})}\right)} \tag{5.1}$$

$$r_{13.24} = \sqrt{\left(1 - \frac{(1 - R^2_{1.234})}{(1 - R^2_{1.24})}\right)} \tag{5.2}$$

where R in the numerator is the coefficient of multiple correlation between X_1 and all three independent variables, and R in the denominator is the coefficient of multiple correlation between X_1 and the two eliminated independent variables.

Applications. Testing the statistical importance of an independent variable in a multiple-regression analysis (M23).

P6   Pascal's triangle

A triangular-shaped table setting out the coefficients in the expansion of the binomial $(x + y)^n$. The following shows the construction of the table up to $n=6$.

Value of n	Coefficients in expansion of $(x + y)^n$
1	1 1
2	1 2 1
3	1 3 3 1
4	1 4 6 4 1
5	1 5 10 10 5 1
6	1 6 15 20 15 6 1

It will be seen that each coefficient from $n=2$ upwards can be derived by adding together the coefficients in the two adjoining columns in the line above. For instance, the third term for $n=4$ is 6. The adjoining columns in the line above have 3 and 3, which together give 6 for the third term for $n=4$. The table can be extended downwards indefinitely. See binomial distribution (B11).

Applications. Quality control.

P7   Passenger mile

A unit of travel by a passenger transport service. Passenger miles are an essential term in a number of important business ratios (B17) specific to a passenger transport undertaking. The ratios are:

$$\text{Average length of journey in miles} = \frac{\text{Passenger miles}}{\text{Number of journeys}}$$

$$\text{Average receipt per mile in £s} = \frac{\text{Passenger receipts in £s}}{\text{Passenger miles}}$$

$$\text{Average load in no. of passengers} = \frac{\text{Passenger miles}}{\text{Vehicle miles}}$$

The average load relates to the train, the bus, the coach or the aircraft, and the train or vehicle miles may include, or exclude, 'empty' running and flying as thought best.

$$\text{Loading coefficient or load factor} = \frac{\text{Passenger miles}}{\text{Seat miles}}$$

Generally the loading coefficient will be less than unity. During rush hours, it probably exceeds unity in large towns and cities, and by an appreciable margin on London Underground and in the Southern Region suburban services of British Rail.

Since the quality of the passenger mile varies from service to service and from place to place, it is less a unit of service than a unit of travel. The ton mile (T10) is the unit of freight transport.

Applications.   In making business ratios for control purposes.

P8   Payback period

The time it takes to recover the outlay on a capital work from the discounted annual yield of the work. If the yield is an annual constant, the formula is

$$\text{Payback period in years} = \frac{\text{Capital outlay}}{\text{Annual yield}}$$

Of all the investment-appraisal techniques used in industry, it is probably the crudest and the most indefensible. It fails to measure the profitability of a work, since, on its own criterion, there can be no profit until the payback period is over, the post-payback yield being entirely ignored.

P9   *P/E* ratio

Of an industrial ordinary or preference share, the ratio of the market price to the earnings per share, or the total market value of the shares to the total earnings of the company (E1).

Before the price-earnings ratio began to appear in the City pages of the newspapers, and before it became a standard figure in the share information service of the *Financial Times,* in the mid-sixties, the two principal share ratios were the earnings-dividend ratio, better known as the *times covered* or *cover,* and the dividend-price ratio, better known as the dividend yield (D21). In effect the *P/E* ratio takes care of both these ratios. Let E represent the earnings per share, D the dividend per share, and P the market price per share, then

$$\frac{1}{E/D \times D/P} = \frac{D}{E} \cdot \frac{P}{D} = \frac{P}{E}$$

Since the dividend yield is usually expressed as a percentage, the numerator in practice is 100 not unity. For instance, according to the *Financial Times* of 18 January 1972, the three ratios for Brooke Bond Liebig 'B' shares at the close of business on 17 January were:

Cover (E/D)   1.9
Dividend yield (D/P)%   3.9
P/E ratio   13.3

$$\frac{100}{1.9 \times 3.9} = 13.5$$

The difference is probably due mainly to rounding (R18).

The *P/E* ratio is a measure of the esteem in which the market holds a share; the higher the ratio, the greater the esteem. It is regarded as indicating potential growth, not necessarily of the company, but rather, of the market price of the share. However, since the current market price of a share can be regarded as standing at a level which fully discounts the market's estimation of potential growth, the dividend yield may be an equally good indicator of the market's esteem.

As the above equation shows, a low dividend yield is consistent with a high *P/E* ratio, but so, too, is a low cover ratio. Before the *P/E* ratio became fashionable, a share was considered to be a good one if it had a high cover ratio. But the fact of the matter is, it still is: despite the inconsistency, nearly all shares with a genuinely high *P/E* ratio have a high cover ratio, the high cover ratio being more than offset by a very low dividend yield. Property shares are an obvious exception to the rule: they mostly have a very high *P/E* ratio, often associated with a moderate cover ratio.

Applications. Investment analysis.

P10 Percentage laws

The percentage rate per unit to which a dependent variable tends to rise or fall when an independent variable is doubled. The eighty per cent law said to apply to learning (L6) is well known: it states that when the cumulative output of a given product doubles, the labour time per unit of output falls to 80%. The laws apply only to relationships that are logarithmic. There are three of them. Let *a* be the regression coefficient (R8) of the logarithm of the independent variable to be doubled, then the three laws are as follows:

Where doubling the independent tends to double the dependent. Let *D* be the percentage law, then
$$D = 50 \times 2^a \tag{10.1}$$
Where doubling the independent tends to halve the dependent. Let *H* be the percentage law, then
$$H = 200 \times 2^a \tag{10.2}$$
Where doubling the independent has no effect on the dependent except through the operation of the law. Let *N* be the percentage law, then
$$N = 100 \times 2^a \tag{10.3}$$
The coefficient *a* should be given its correct sign, which is that shown in the regression equation. Where (10.1) applies, the sign of *a* is invariably plus, where (10.2) applies, it is invariably minus, and where (10.3) applies, it may be either plus or minus. To give percentage laws of 100, the value of *a* in (10.1) is +1, in (10.2) it is −1, and in (10.3) it is 0.

Applications. Presenting to non-mathematicians the results of a logarithmic regression analysis.

P11   Percentile

Any of the 99 values that divide a frequency distribution into 100 equal parts. The 10th percentile is the first decile (D1), the 25th is the first quartile (Q5), and the 50th is the median (M14).

Applications. Analysing frequency distributions.

P12   Perfect competition

Exists where market forces alone decide the price. Even governments cannot exercise control of prices, where competition is perfect, without first making it imperfect by rationing or the control of supplies. In order to have control of their own prices, suppliers have sought to create market imperfection by branding (B13) and by forming cartels and trusts.

P13   Permutations

The arrangements that can be made by taking some, or all, of a number of things. Each arrangement is called a *permutation*. The number of permutations that can be made from *n* objects taken *r* at a time is

$$\frac{n!}{(n-r)!}$$

In race-course jargon, a dual forecast requires the first and second horse to be named in order. If a punter selects five horses as potential winners of a race, he can make 20 permutations taking two at a time, ie

$$\frac{5 \times 4 \times 3 \times 2}{3 \times 2} = 20$$

If we distinguish the five horses by the letters, *a, b, c, d, e,* we have the following permutations

ab ac ad ae	(and reverse) give 8
bc bd be	(and reverse) give 6
cd ce	(and reverse) give 4
de	(and reverse) give 2
	Total 20

The formula is that for combinations (C15) with factorial *r* as a separate term excluded from the denominator, ie the number of permutations is factorial *r* times the number of combinations of *n* things taken *r* at a time.

Applications. General use, critical path or network analysis.

P14 Pictogram
A picturegram. A chart giving statistics in pictorial form.

Applications. Popular presentation.

P15 Pie chart
Sometimes referred to as a *cake diagram,* a chart in which some figure is broken down into its constituent parts. A number of degrees out of the total of 360° is allocated proportionally to the total to each part. If the total figure is £600, then a part consisting of £100 would be allocated one-sixth of 360° = 60°. The area of the slice and the ·corresponding arc on the circumference are also proportional to the total.

Applications. Presentation.

P16 Pilot survey
A preliminary sampling inquiry, usually very small, carried out to indicate the best design to adopt for a full inquiry.

Applications. Collection and collation of new statistics; preparing questionnaires, sampling.

P17 Plant capacity
The capacity of a machine, a factory or a vehicle expressed in terms of the rate of output of goods or services under given operating conditions. There are two measures, both somewhat theoretical in concept, but both important in practice: (1) normal capacity; (2) economic capacity. Normal capacity can be defined as the maximum rate of output achievable before the marginal cost begins to increase as a result of the special measures that have to be taken to increase the rate of output further. Economic capacity can be defined as the rate of output at which the marginal cost is equal to the unit cost.

Table M4.1 contains an example of output whose rate rises above the normal capacity of the plant and reaches the economic capacity. Figure P17.1 expresses the idea graphically. It may pay to increase the rate of output above the normal capacity up to, but not above, the economic capacity. When the rate rises above the normal capacity, then, by definition, there is an increase in the marginal cost, which provides a term in the optimum-price formula. See marginal cost (M4).

In the example of Table M4.1 the marginal cost is shown as ranging from £3.0 for rates of output up to the normal capacity, to £3.94 for a rate of output equal to the economic capacity. If the

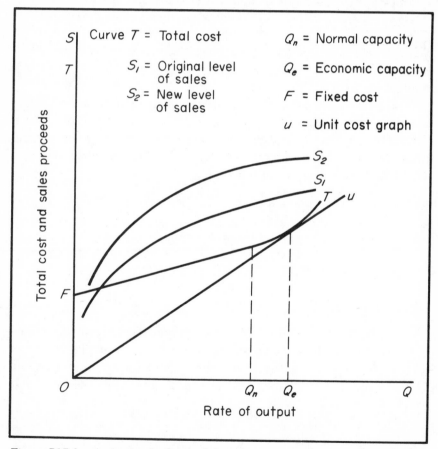

Figure P17.1   A rise in the level of the sales curve, and a rate of output that rises above the normal capacity of the plant

elasticity of demand for the product is 2, then the optimum price
(R6) is

$$p_m = 3\frac{2}{2-1} = £6$$

for rates of output and sales up to the normal capacity, and
apparently

$$p_m = 3.94\frac{2}{2-1} = £7.88$$

for a rate of output equal to the economic capacity. It will be seen
from the demand equation (D7.1) that provided the elasticity of
demand remains constant, an increase in the level of demand at the
optimum price results in an increase in the value of the constant K.
When this happens, the sales curve swings upwards to a new position.
In Figure P17.1, the original position of the sales curve is marked S_1,
and the new position, S_2. The problem now is whether it would be
more profitable to charge the higher price of £7.88, and so damp
down the actual demand (which would not cause or prevent any
movement of the sales curve), or to extend the plant, thus incurring
additional annual capital charges, ie fixed costs, and so raise the total
cost curve to a new position, and to continue charging the original
price of £6. If the original value of K in the sales equation is
1 080 000, then the optimum sales are

$$Q_m = \frac{1\,080\,000}{6^2} = 30\,000 \text{ units}$$

which is equal to the normal capacity of the plant.

Suppose now the value of K rises to 1 260 000, then the sales at a
selling price of £6 would be

$$Q = \frac{1\,260\,000}{6^2} = 35\,000 \text{ units}$$

which happens to be equal to the economic capacity of the plant.
But suppose also that the company knows that the marginal cost has
risen to £3.94 — which, as shown above, gives an optimum price of
£7.88 — and decides to increase the price accordingly. Then sales fall
to

$$Q = \frac{1\,260\,000}{7\,88^2} = \frac{1\,260\,000}{62.1}$$

$$= 20\,290 \text{ units}$$

But this is much lower than the *normal* capacity of the plant, so that it can be said that £7.88 is a suboptimum price. With the new level of demand, the true optimum price would dispose of at least 30 000 units a year:

$$p^2 = \frac{1\,260\,000}{30\,000} = 42$$

Therefore $p = £6.48$

Table P17.1 Net-revenue schedule for a case where actual demand rises above the normal capacity.

	Annual sales		Annual cost	Net revenue
Price	Quantity	Value		
p	*Q*	*pQ*		
(£)	(no)	(£'000)	(£'000)	(£'000)
6.00	35 000	210.0	137.6	72.4
6.48	30 000	194.4	120.0	74.4
7.00	25 720	180.0	107.2	72.8
7.88	20 290	159.9	90.7	69.2

Table P17.1 contains an abbreviated net-revenue schedule based on the cost schedule of Table M4.1, with gross revenues mostly calculated by reference to the above 'optimum' prices. It shows, or at least indicates that under conditions of the higher level of demand and the existing capacity of the plant, the optimum price in this case is that which equates the actual demand to the normal capacity of the plant, ie £6.48. The annual cost figures in the table for rates of output not shown in Table M4.1 are below the normal capacity and are calculated by taking the annual output at the marginal cost of £3.00, and adding a fixed cost of £30 000 a year.

Applications. Investment programming, investment appraisal, efficiency auditing.

P18   Point-of-time statistics
Statistics that relate to a point of time as distinct from rate-of-flow statistics. Point-of-time statistics include stock figures, the book values of fixed assets (F5), price, marginal cost (M4), marginal

revenue (M6) and unit cost; rate-of-flow statistics cover figures that are expressed per unit of time (day, week, year, etc.), such as operating costs, sales, annual costs and fixed costs.

P19   Poisson distribution

A distribution which generally concerns unrelated events, where neither the binomial (B9) nor the normal (N9) distribution seems to apply. It exploits the exceptional properties of e = 2.7183, (to four decimal places), the base of natural or Napierian logarithms. It is equal to the sum of a series of figures resembling a frequency distribution,

$$e = 2.7183 = \frac{1}{0!} + \frac{1}{1!} + \frac{1}{2!} + \frac{1}{3!} + \dots \tag{19.1}$$
$$= 1 + 1 + 0.5 + 0.17$$

If e is raised to the power m, e^m, then all the terms of its expansion are similarly raised, and the first two terms are no longer identical, except where $m = 1$. Where m is less than 1, the first term is greater than the second term, and where it is greater, the first term is smaller.

If each term of the expansion is divided by e^m (or, as it is usually expressed in the literature, multiplied by e^{-m}, the sum of the terms is equal to unity:

$$1 = \frac{1}{2.7183^m} + \frac{m}{2.7183^m} + \frac{m^2}{2.7183^m 2!} + \frac{m^3}{2.7183^m 3!} + \dots$$

Assuming m is the expectation, it can be determined by dividing the total number of observations into the total number of times that the event occurs.

It is suggested that anyone intending to apply the Poisson distribution as a regular routine would be well advised to draw up a table of the value of $1/e^m$ for values of m, ranging from $m = 0$ to $m = 5$, on the lines of Table P19.1, which provides the first term of the distribution, and the product of the two, the second term.

A simple numerical example of a percentage distribution can be used to demonstrate it. The example used (Table P19.2) is that of the percentage distribution of housewives watching ITV during a week in November 1972, published in an article on 'What's in it for the advertiser' by David Wheeler in the *Financial Times* of 25 January 1973.

Table P19.1   Values of $1/e^m$ and $m(1/e^m)$.

m	$1/e^m$ (1st term)	$m(1/e^m)$ (2nd term)
0.000	1.0000	0.0000
0.005	0.9950	0.0050
0.010	0.9901	0.0099
0.015	0.9662	0.0145
0.020	0.9551	0.0191
0.050	0.9512	0.0471
0.100	0.9048	0.0905
. . .	. . .	. . .
4.800	0.0082	0.0394
4.900	0.0074	0.0363
5.000	0.0067	0.0335

It often happens in practice that m is difficult to determine. This is one of those cases. Within the known limit of up to 25 hours watching, there are five terms each covering a range of five hours, plus one term, that for zero, so that we have 5 x 5 plus 1 = 26. The percentage of housewives spending up to 25 hours in the week watching ITV was 78. It is suggested that m equals this figure divided by 26, ie $m = 3$. We then have $e^{-m} = 2.7183^{-3} = 0.0498$. Table P19.3 gives the Poisson distribution for $m = 3$, together with the actual sample figures for November 1972 (also for November 1971).

Table P19.2   Weekly ITV viewing by housewives.

Hours spent	%
0	7
0-5	16
5-10	17
10-15	15
15-20	13
20-25	10
Over 25	22
	100

Table P19.3 Poisson distribution for housewives' ITV viewing.

Hours spent (number)	Poisson distribution (per unit)	(%)	Actual distribution November 1972 (%)	November 1971 (%)
0	0.0498	5.0	7	4
0-5	0.1494	14.9	16	18
5-10	0.2241	22.4	17	20
10-15	0.2241	22.4	15	18
15-20	0.1681	16.8	13	13
20-25	0.1009	10.1	10	11
25-30	0.0504	5.0		
30-35	0.0216	2.2		
35-40	0.0081	0.8	22	16
40-45	0.0026	0.3		
Residue *	0.0010	0.1		
Total	1.0000	100.0	100	100

* Probably due to watching in excess of 45 hours a week, rather than to rounding.

Although the Poisson distribution is based on 1972, it seems to provide a better fit to the 1971 sample figures. For November 1971, the value of m achieved in the way described above for 1972 is 3.2. The first few terms of the Poisson distribution for $m = 3.2$ are

Hours spent (number)	Poisson distribution (Per unit)	(%)	Actual distribution Nov 1971 (%)
0	0.0408	4.1	4
0-5	0.1232	12.3	18
5-10	0.2089	20.9	20
10-15	0.2228	22.3	18
15-20	0.1783	17.8	13

The χ^2 test (C12) can be applied to each term to test it for goodness of fit.

Applications. Quality control, market research; extending frequency distributions.

P20   Population forecasts

Of all UK official statistics, those of population are the only ones that are projected officially into the future for more than 18 months ahead. The forecasts include analyses by country (England, Wales, etc.), by sex and by age group, and give estimates for each year up to 1976, and then decennially from 1981 to 2011. The forecasts are based on assumed death-rates and birth-rates, projected from the figures collected at the Census of 1971, and are prepared by the Government Actuary's Department.

Sources. Census of Population Reports; Annual abstract of Statistics, both HMSO.

Applications. Sales forecasting, business forecasting generally.

P21   Positive correlation

Exists statistically where, in a comparison of two cross-section series (C37) or time series (T8), high figures in one are associated with high figures in the other, and low with low. Other things being equal, the statistics of annual advertising expenditure are positively correlated with those of annual sales turnover; the same applies to statistics of cost and output, of profits and turnover, and productivity and expenditure on labour-saving devices. However, as the entry on correlation (C25) shows, statistical correlation is sometimes misleading owing to the intervention of a third factor.

P22   Premature displacement of fixed assets

The displacement of wasting tangible fixed assets (F5) before they reach the end of their book lives, or before they are worn out. The disposal of freehold land and earthworks, surplus to the firm's requirements, or of the goodwill in a business can scarcely be called *premature displacement.* There are three potential causes of premature displacement:

1   Technological progress:
a   *Internal.* The development of machinery and plant designed to save labour, rendering earlier types obsolete.
b   *External.* The development of domestic appliances which render earlier types of appliances obsolete.
2   Changing fashion and consumer taste, which is entirely external, but which may render some types of machinery and plant obsolete.

Of these, by far the most important is 1a. This means that most asset displacements are made under renewal schemes, ie the assets are displaced to give room for new assets designed to perform the same tasks more efficiently.

A problem that arises in these circumstances concerns the financial aspect: the total provision made in respect of the depreciation of a prematurely displaced asset is not enough either to amortize the original outlay on it or to renew it, like for like. How should the deficit be financed? The question has given rise to some controversy. There are two schools of thought. One argues that the deficit should be written off — it is only a book-keeping transaction, so that the idea of a deficit is unrealistic. The other argues that if a depreciation charge is worthwhile, it is more than a book-keeping transaction and the deficit should be financed out of the profit of the new technology which caused the premature displacement. The latter school appears to have the more logical argument. How the deficit would be notionally financed in the way suggested at the investment programming stage is discussed under unprovided renewals (U4). for depreciation, see D10; for replacement of fixed assets, R11.

P23   Present value (PV)

The value at the beginning of a period of years of a sum of money payable at some date years hence, or of sums of money receivable at different dates. For purposes of calculation, there are two distinct formulae:

1   The PV of £ payable in n year's time at r rate of interest.
2   The PV of £ a year for n years at r rate of interest.

The present value is always smaller than the amount, or sum of the amounts payable, entirely on account of interest, (ie not on account of inflation, though admittedly the formulae can be used for calculating real values as well as money values (M19). Here, we are concerned with money values.) We examine the two concepts separately:

PV of £ in n *years* is equal to the reciprocal of the amount of £ in n years (A8), and the algebraic formula is therefore

$$PV \text{ of } £ = (1 + r)^{-n} \qquad (23.1)$$

where r is the rate of interest per £, and n is the number of years. A sum of £100 invested now at 10% will amount to £110 at the end of one year. The present value of £110 payable a year hence is therefore £100 with interest at 10%. It follows that the PV of £100 payable a year hence is 100/110 of £100 = £90.9091 to six significant figures, ie the PV of £ is the reciprocal of the amount of £.

PV of £ a year for n *years* is equal to the sum of the present values of £ in each year up to the *n*th. The direct algebraic formula is

$$\text{PV of £ a year for } n \text{ years} = \frac{1-(1+r)^{-n}}{r} \qquad (23.2)$$

The PV of a perpetuity, ie of £ a year in perpetuity, is equal to the amount that would have to be invested in an irredeemable security to yield an annual amount equal to the perpetuity, eg £1000 is the PV of a perpetuity of £100 a year with interest at 10%.

For the PV of various sums payable in various years, it is necessary to take the total of the PVs of the several sums calculated separately by (23.1) above.

For the present value of £ a year for *n* years deferred *N* years, see deferred annuity (D5).

Applications. Investment appraisal, effiency auditing; actuarial calculations for pension funds.

P24 Price discrimination

The charging of different prices to different customers; usually results from differences in size of order and delivery, and often takes the form of a discount for large orders. It may also result from differences in the distance of purchaser's premises from the supplier's, when the supplier provides transport or pays the carriage charges. Different ex-factory or ex-warehouse prices are nearly always difficult to justify, especially where production is continuous. If one ex-works price of the firm is the optimum (R6), any other in the same market is likely to be a suboptimum.

P25 Price index

The ratio of the average of prices in one period or place (referred to as the *given period* or *place*) to the average in another period or place (referred to as the *base period*). The subject is fully dealt with in the entries listed under I7.

P26 Price relative

The ratio of the price of an article or service in one period or place, to the price in another:

$$\text{Price relative} = \frac{p_n}{p_0}$$

where p is the price, and n and 0 denote the given and base periods or places, respectively. An essential ingredient in practical index-number formulae. See index numbers, I7.

Applications. Making price-index numbers.

P27 Price under perfect competition

A price decided by supply and demand in the market, and over which the supplier has no control. Perfect competition exists in auction markets such as those dealing in cattle, fruit, vegetables, corn, metals and other raw materials, and on the stock exchange.

P28 Probability

Likelihood; in statistics, its measure ranges from zero for absolute impossibility to 1.00 for absolute certainty, though admittedly, both are rarely seen in practice. It is usually symbolized by p. When $p = 0.5$, there is a 50-50 chance of an event occurring, ie in race-course jargon, the odds are even. Sometimes the probability is expressed in percentage form, 100% being absolute certainty. If a coin is tossed, there is a probability of 100% that it will turn up either heads or tails, 50% that heads will turn up, and zero probability that neither will turn up.

There are several circumstances for which a measure of the probability is often required. These are examined in the following entries:

A5   Addition law
B9   Binomial distribution
C12   Chi-squared test
F9   Frequency distribution
F10   *F* test
M24   Multiplication law
N9   Normal curve of error
P19   Poisson distribution
P30   Probable error

S3　　Sampling
S25　　Standard error
S30　　Statistical quality control
S37　　Student's *t*

P29　Probability paper
A type of graph paper ruled to show ranges of probability and expectation and different numbers of occurrences of an event; used in lieu of tables.

Applications.　Statistical analysis, quality control.

P30　Probable error
Equal to 0.6745 of the standard error (S25); the measure of the error of a mean of a large sample (more than 30 observations) where the chances are even that the mean of the universe lies within it. The concept of the error with a 50% probability has fallen into disuse in favour of one with a 68% probability, equal to one standard error. See normal curve of error (N9).

P31　Production function
Expresses annual output as a function of the average numbers employed, the cumulative output, the average size of the product in terms of weight, volume, area or length, and any other relevant factor. The evidence available suggests that the form of the model is logarithmic:
$$\log Q = \log K + a \log F + b \log W + C \log E \qquad (31.1)$$
where Q is the annual rate of output, K a constant, F the average labour force, W the average size of product, E cumulative output of the product, and a, b and c are the coefficients. See employment function (E5), also L6, P10 and S7.

Applications.　As the model for a regression analysis of production.

P32　Production statics
Consist for the most part of sales and deliveries ex-works, in terms of both quantity and value. The official monthly index of production is derived largely by dividing the value index by the price index for each sector. Since the price index numbers are Laspeyres's (L4), the production index numbers thus derived are Paasche's (P1), as the

entry on index numbers (I7) shows.

Sources. *Census of production reports, Monthly Digest of Statistics, Annual Abstract of Statistics, Business Monitor* (four series — production, census, service and distributive, and miscellaneous — of regular reports on over 150 industries, available on subscription), *Abstract of Regional Statistics, Scottish digest, Welsh Digest, Northern Ireland Digest, Ministry of Power Statistical Digest, Agricultural Statistics of the United Kingdom, Housing Statistics for Great Britain* (all published by HMSO).

For iron and steel production: *Monthly Bulletin* and *Year Book* of the British Iron and Steel Federation and the Steel Board.

For non-ferrous metals: *World Statistics of Non-ferrous Metals.*

For wool: *Monthly Bulletin of Statistics* (Wool Industry Bureau of Statistics).

For cotton: *Cotton Board Quarterly Statistical Review*

For rubber: *Rubber Statistical Bulletin* (International Rubber Study Group).

Some trade associations also collect and publish production, etc. statistics. An example is the *Monthly Statistical Review* published by the Society of Motor Manufacturers and Traders, who also publish a *Year Book* containing annual series of statistics.

P33   Productivity

Unqualified, the output per man hour, man month or man year. Where it refers to a factor of production other than labour, it is usually qualified, eg the productivity of land under wheat in 1971 was 35cwt per acre. The productivity of capital is somewhat ambiguous, and where it refers to the output of a particular type of machine or vehicle, it is probably best expressed in terms of labour productivity.

It is believed in some quarters, that the use of value added (V1) as the numerator in compiling a productivity index number (I7) would have the effect of eliminating from the index the unwanted changes due to changes in make-or-buy policy. Value added gives a biased index of productivity. The reasoning behind this statement is: when a change in make or buy takes place, then, other things being equal, the wages bill rises or falls *pro rata* to the labour employed. Since the wages bill forms only part of the value added, value added is bound to rise or fall proportionally less than the labour force. This means that, other things being equal, a change from make to buy would of itself cause a rise in the index of productivity, and a change from buy to make would cause a fall.

It seems to follow that a productivity index for a company must have as its numerator a production index built up in detail with new items (eg buy-to-make items) spliced on, and discarded items (eg make-to-buy items) spliced off. (The process of splicing on and off is described in most textbooks, eg E. J. Broster, *Planning Profit Strategies,* Longman, 1971, pp. 249-51.)

A production index has the additional attraction over value added of being applicable to individual processes and products, as well as the factory or firm as a whole; this renders the resulting productivity index numbers of greater value in productivity bargaining and the like.

A conventional production index weighted by values would scarcely meet the needs of a productivity index. A labour-weighted index of production seems to be appropriate. An example (taken from E. J. Broster, *Management Statistics,* Longman, 1972) can be used for demonstrating the calculations.

Suppose a jam factory has three main processes:

1   Making, including preparing fruit and boiling.
2   Filling into pots each containing 1 lb of jam.
3   Capping the pots, labelling and packaging for distribution.

In 1969, the factory makes, fills, caps, etc. 100 000 lb of jam. In 1970, advanced filling and capping technologies are introduced, with considerable savings in manpower in 1971. There is no difficulty in ascertaining the productivity at the process level. If the factory makes 150 000 lb in 1971, and the labour force increases from 40 men in 1969 to 60 men in 1971, the productivity in the making department remains the same. If the labour force averages 50 men in 1971, then there is an increase in productivity of 20%, the index number for 1971 being 120 to the base 1969 = 100:

$$\frac{150}{100} \cdot \frac{40}{50} = 1.20$$

Suppose the firm has overestimated the demand in 1971 and, although 150 000 lb of jam are made and filled into pots, the factory caps, labels, etc. and sells only 120 000 lb in that year. Table P33.1

gives the detail and the calculations needed to provide an overall labour-weighted production index for the factory. The ratios in the last column of the table provide the productivity index numbers, process by process.

Table P33.1   A labour-weighted production index.

	1969			1971				
	Output lb ('000)	Man years (no.)	Man years (per 1000 lb)	Output lb ('000)	Man years (no.)	Man years (per 1000 lb)		$\dfrac{m_0}{m_n}$
Process	q_0	$m_0 q_0$	m_0	q_n	$m_n q_n$	m_n	$m_0 q_n$	
1	100	40	0.400	150	50.0	0.3333	60.0	1.20
2	100	10	0.100	150	6.0	0.0400	15.0	2.50
3	100	3	0.030	120	1.5	0.0125	3.6	2.40
Total (Σ)		53			57.5		78.6	

$$\text{Production index } Q_{n.0} = \frac{\Sigma m_0 q_n}{\Sigma m_0 q_0} = \frac{78.6}{53.0} = 1.483$$

For the three processes of Table P33.1, the appropriate labour ratio is $\Sigma m_n q_n / \Sigma m_0 q_0$, which corresponds to the value ratio, in which p serves for m. For the example, the ratio is 57.5/53.0 = 1.085, which is divided into the production index to give the productivity index. However, if the productivity index for the factory as a whole is required, including warehousing labour and office staff, the total labour forces in the given and base periods would be used for arriving at the labour ratio to use as the denominator.

Let the productivity index in the given year to 1.00 in the base year be represented by $M_{n.0}$, then, for the three departments or processes:

$$M_{n.0} = \frac{1.483}{1.085} = 1.367$$

However, since the ratio m_0/m_n gives the productivity index, process by process, one might expect that, weighted for output, it would give a productivity index for the three processes combined. And so it does. The index derived from the weighted figures is

$$M_{n.0} = \frac{\Sigma m_0 q_n}{\Sigma m_n q_n} = \frac{78.6}{57.5} = 1.367$$

which is exactly the same as the figure derived above by dividing the labour ratio into the production index.

We have the labour-weighted production index (see Table P33.1)

$$Q_{n.0} = \frac{\Sigma m_0 q_n}{\Sigma m_0 q_0} \tag{33.1}$$

and the labour ratio

$$L_{n.0} = \frac{\Sigma m_n q_n}{\Sigma m_0 q_0} \tag{33.2}$$

Equation (33.1) divided by (32.2) gives

$$M_{n.0} = Q_{n.0}/L_{n.0} = \frac{\Sigma m_0 q_n}{\Sigma m_0 q_0} \cdot \frac{\Sigma m_0 q_0}{\Sigma m_n q_n} \tag{33.3}$$

The term $\Sigma m_0 q_0$ cancels out, leaving

$$M_{n.0} = \frac{\Sigma m_0 q_n}{\Sigma m_n q_n} \tag{33.4}$$

which is the same as the above; it is, in effect, the inverse of Paasche's price index weighted for quantity, with m serving for p.

However, (33.4) is limited in its application to the derivation of productivity index numbers in respect of the labour force accounted for in the process calculations. If, as suggested above, an overall productivity index for the factory is required, then it would be necessary to apply (33.3), in which the labour ratio is the total factory labour force in the base period divided by that in the given period. If, for our hypothetical jam factory, the two labour figures are 90 and 100, respectively, the overall productivity index would be 90/100 of 1.483 = 1.335.

The example introduces another problem, that of a case in which a product is stored awaiting final processing and delivery in the following year. Value added could not take care of the labour expended on the surplus going into store, even if it were otherwise satisfactory in providing a valid numerator for an overall index for the factory or firm.

Applications. Production cost control, productivity bargaining.

P34   Product mix
The body of ratios of the various products made or sold by a company to the total in terms of value. *Optimum product mix* is that body of ratios which maximizes the company's profit. It is a technique of linear programming (L10) but, as usually expounded, it makes a number of unrealistic assumptions, eg that the company's assets can be switched from producing one type of article to producing any other of the company's range, and that there is no limit to the demand for any product at a fixed price. See break-even analysis (B14) and rational price fixing (R6).

P35   Product moment
The product of the deviation from the mean of a value of X and that of the corresponding value of Y.

P36   Profit planning
Any planning designed to increase or at least maintain the company's profits could be described as profit planning. However, in the literature it has come to mean an extension of linear break-even analysis (B14).

Let P represent profit, S sales proceeds, T total cost of sales (C31), V variable cost, F fixed cost, and C contribution, all in unit time, say, a year. Then

$$C = S - V = F + P$$
$$P = S - T = S - (V + F) = C - F$$

in a break-even situation we have
$$P = 0 \text{ and } S = T$$
$$F = S - V = C$$
$$SF = S(S - V) = SC$$

$$S = \frac{SF}{C} = F\frac{S}{C} \qquad (36.1)$$

In the entry on implied elasticity of demand (I2), Equation (I2.4) reads:

$$e_i = \frac{p}{p - a}$$

where e_i is the implied elasticity of demand, p the price, and a the marginal cost. Multiplying the numerator and denominator by Q, the quantity sold, we have:

$$e_i = \frac{pQ}{Q(p - a)} = \frac{S}{S - V} = \frac{S}{C} \qquad (36.2)$$

The term S/C in (36.1) is thus equal to the implied elasticity of demand, and if the fixed price of the linear analysis happens to be the optimum (R6), it is equal to the true elasticity of demand. Equation (36.1) may therefore be rewritten:

$$S = Fe_i \qquad (36.3)$$

which gives the value of sales when $P = 0$.

Linear analysts speak of the P/V ratio, which is defined as the 'profit/volume ratio', ie

$$\frac{100(S - V)}{S} = 100\frac{C}{S} \qquad (36.4)$$

It will be seen that 'profit' is equal to the sales proceeds less the variable cost, and V is the sales proceeds (not the volume or quantity of sales). How the ratio came to be called the P/V ratio is unclear: the 'percentage C/S ratio' would make more sense, since it conforms to the linear analysts' notation, in which (the P/V ratio apart) P means net profit (ie $S - T$), and V the variable cost.

Nevertheless the ratio is of practical value. It provides a measure of the annual contribution to profits and fixed costs (which in the circumstances is equal to net profit) that each increase in sales of £100 year will make. If the selling price is the optimum, the additional contribution is equal to £100 divided by the elasticity of demand, ie

$$P/V \text{ ratio} = 100\frac{C}{S} = \frac{100}{e} \qquad (36.5)$$

If S_y is the break-even sales, and S the actual sales, then $S - S_y$ can be called the margin of safety. The contribution or net profit made by the product is equal to

$$\frac{(S - S_y)C}{S} = \frac{S - S_y}{e_i} \text{ or } \frac{S - S_y}{e} \qquad (36.6)$$

as the case may be.

Where Q_n represents the normal annual capacity of the plant (P17), then, at the fixed price p, the maximum net profit to be derived from the product is

$$\frac{pQ_n - S_y}{e_i} \qquad (36.7)$$

or

$$\frac{p_m Q_n - S_y}{e} \qquad (36.8)$$

However, it is shown in P17, that, where output exceeds the normal capacity of the plant, the optimum price is close to that which equates demand to the normal capacity, a price somewhat in excess of the optimum for rates of output within the normal capacity of the plant. If (36.8) is applied to such a situation, the value of the new optimum price in the first term of the numerator is greater than the optimum price on which S_y is based. The break-even quantity may be written:

$$Q_y = \frac{F}{p - a}$$

where $p - a$ is the unit contribution. It will be seen that, since a is a constant (see M4) for rates of output within the normal capacity of the plant, the value of Q_y changes with variations in p, and so, therefore, does the break-even variable cost.

A problem that must be faced in this kind of analysis is the definition of fixed cost. Should it be the attributable fixed cost (A15) or the attributable fixed cost plus arbitrarily allocated costs? The complete absence of fixed costs would not make nonsense of profit planning. Graphically depicted, the break-even point would lie at the origin, and $Fe = FS/C = 0$; the P/V ratio so called would be

$$\frac{100(S - T)}{S}$$

and the margin of safety would be S/e_i or S/e.

Applications. Financial control of individual products, investment appraisal and programming, efficiency auditing.

P37   Profit variance
The extent to which the profit in a period differs from the budgeted profit see A10, B15.

Applications.   Budgeting, net-revenue analysis.

P38   Public opinion poll
A sampling inquiry (S3) carried out to test public opinion on some question of the day, or on some commercial product as part of a market-research exercise. Public opinion polls on questions of the day are conducted by quota sampling (Q9) organized to save time rather than expense. Field workers are employed for the purpose, and each interviews his quota of people (about 40-50 in number) in the streets and other public places. The operators have to depend on their field workers not to load their questions. In the circumstances, bias (B8) cannot be avoided but the operators try to minimize it by taking large samples. See R5, S5, U3.

Applications.   **Market research.**

P39   Purchasing power of money
The quantity of goods that a unit of money will purchase, usually expressed in index number form (I7). To the community in general, it is the inverse of the retail-price index (C30); to a manufacturer, it may be the inverse of the price index of the materials he purchases and the labour he employs (M19).

P40   *P/V* ratio
See Profit planning (P36).

q

Q1 Quadrant

One of the four quadrants of the graph (G5), as depicted in Figure Q1.1. Each quadrant is identified by number or its compass position as shown. The NE quadrant is no. 1 and is the one that is most used, as it accommodates the positive values of X and Y:

Sign of	Quadrant	
X	Y	
+	+	No. 1: NE
−	+	No. 2: NW
−	−	No. 3: SW
+	−	No. 4: SE

The horizontal axis is called the x-axis, and the vertical axis, the y-axis. Their point of intersection is called the origin, and is marked 'O': it provides the position of the dot for $Y = X = 0$.

Q2 Quadratics

Equations involving X^2 eg

$$Y = aX^2$$
$$Y = aX^2 + K$$
$$Y = aX^2 + bX + K$$

Rarely found by deduction in the business sphere, but may exist empirically. A quadratic should never be adopted as a form of management model until careful consideration has been given to its implications. Finite differences provides a technique for obtaining empirical confirmation of a quadratic or higher-degree relationship: it is demonstrated in F3.

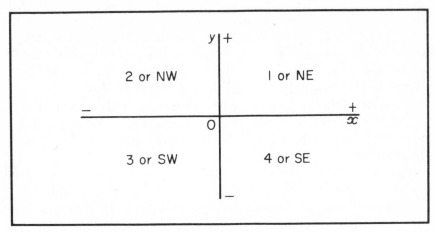

Figure Q1.1   The four quadrants of the graph

Q3   Quality control

The systematic control of the quality of goods purchased, or goods to be sold, by the inspection and analysis of random samples (R3). Since World War I, quality control has reached a high degree of mathematical sophistication. Briefly, standards are set by the supplier or the purchaser, or both in collaboration. The standard may consist of a defective rate of, say, 3% with a probability of, say, 0.95 that this rate will not be exceeded. Or tolerances may be set with, say, a 0.98 probability that the product will conform to them.

Accurate inspection is all-important; in many cases this means accurate measurement of diameters of holes, widths of rods, lengths, breadths and widths of articles, weights, lives of electric lamps, electronic valves and a wide range of products such as army boots, transmission belting and chains, ball and roller bearings and lubricants. For lives, there is usually only one tolerance limit, the lower one. It is important to the maker's reputation that his electric lamps should have an average burning life of, say, 1000 hours, with a lower limit of 900 hours. To set an upper limit would serve no useful purpose.

The methods of inspection and measurement of items in a sample are now almost as sophisticated as the methods of analysing the sampling data. A well-known method is the go/not-go method of inspection. In this, measurement is not undertaken, and only acceptance or rejection is decided. For inspecting the size of a hole in a piece of metal for instance, a plug is used, thin at one end and thick at the other, the measure of the former being the lower control limit or tolerance, and that of the latter, the upper limit. If the plug cannot be inserted in the hole, the hole is too small; if the plug passes entirely through the hole, it is too large, and the piece is rejected. For inspecting the width of an axle, two rings, one small and the other large, are used in the same way − *go* through the larger ring, *not go* through the smaller ring to pass inspection.

However, under modern principles of quality control, analysts prefer to have the average measure of each sample, its standard deviation (S24), its sampling error (S4), and its range ie the difference between the largest and smallest items in the sample. Although go/not-go plug and ring gauges may still have an important part to play in inspection, in some cases the analyst needs to have an accurate measurement of each item in order to make his calculations and prepare his control charts.

Control on the basis of acceptance or rejection is called control sampling by attributes, and control on the basis of accurate measurement is called control sampling by variables. The approach to one is different from that to the other, but in both, a number of samples forms the basis of the analysis, and a new sample is drawn from time to time to make sure than the manufacturing process remains reasonably stable.

Sampling by attributes. If, in the samples so far drawn, covering 1000 items, it was found that 20 were defective, we have 0.02 as the proportion. We would therefore expect that any future sample would have a total defective of $0.02n$, where n is the number in the sample, and a total accepted of $0.98n$. This gives a binomial (B11) distribution $(0.98 + 0.02)^n$, from the terms of the expansion of which it would be possible to determine the probability of having 0, 1, 2, 3 . . . rejections in our new sample. A binomial whose two terms are so different has a very skew distribution, and the Poisson distribution (P19) may therefore prove to be good enough for the purpose.

In a sample of 50 items, we would expect one defective, and in one of 100 items, two defectives, but the number will range from 0 upwards, and it is the greatest possible number permissible we seek to establish. Any number of defectives greater than this must lead to immediate action. First, another sample must be drawn and inspected and, if it confirms the results of the last sample, then the only conclusion is that something has gone wrong in the production department. It should be mentioned that the calculated greatest permissible number applies to a predetermined probability. It may be the number that could be expected to occur once in 500 samples or once in 1000 samples.

The control chart for defectives has the sample identification number marked on the x-axis (X1) at equal, but conveniently arbitrary distances apart, and the number defective on the y-axis (Y1). The x-axis serves as the base line for zero defectives. Two horizontal lines may be drawn across the chart, the lower one to represent a warning limit, and the upper one an action limit. The samples should be all of the same size.

Sampling by variables. For this, there are usually two control charts, one the *mean chart,* the other the *range chart.* For both, all samples should be of equal size; the sample identification number is marked along the x-axis. In the former the y-axis provides the measure of the averages of samples, and in the latter, the measure of the ranges of samples.

A horizontal line is drawn across the mean chart to represent the grand average, ie the overall average of the averages of the original series of samples. Then the average of each new sample is plotted on the chart. One horizontal line above the average line is drawn to represent the upper control limit, and another below, to represent the lower control limit. The distance between the average line and the control limit lines depends upon the average range of the original samples. The average range multiplied by a factor called the *A factor,* which is 1.23 for samples of two items ranging to 0.20 for samples of 10 items, is added to the average of the original sample averages to provide the upper control limit and subtracted to provide the lower control limit. Any sample average falling outside these limits would call for action, especially if the range of the same sample falls outside the control limits of the range chart.

A horizontal line is drawn across the range chart to represent the grand average range, thus providing a guide line for the ranges of all new samples which are plotted in the chart. Control limits are also horizontal lines, and their distances from the grand average range depends on the size of the samples and the grand average range multiplied by a factor called the *D factor,* which ranges from 2.9 for samples of two items each to 1.6 for samples of 10 items, for the upper control limit, and from 0.05 for samples of two items to 0.5 for samples of 10 items, for the lower control limit.

As it is common practice in continuous and regular sampling for quality-control purposes to use small samples of four or five items each, it is worth mentioning that the values of the F factor and D factor for these sample sizes are

Sample size in items:	4	5
A factor	0.476	0.377
D factor, upper limit	1.93	0.29
lower limit	1.81	0.37

Calculating the standard deviation, which is useful in quality control, can be a long and laborious task when the number of items in the original set of samples may exceed 1000 by a considerable margin. For sample sizes of three to nine items each, a short-cut approximation to it may be used. It is equal to the grand average range divided by the square root of the number of items in each sample.

When a new system of statistical quality control is being considered, it is advisable to experiment with it alongside the old method before finally introducing it.

Q4   Quantity relative
The ratio of the physical quantity produced, sold or in stock of an article in a given period, or at a given time or place, to the quantity in a base period, time or place, ie

$$\text{Quantity relative} = \frac{q_n}{q_0}$$

It is used in the making of quantity index numbers (I7), a quantity index number being the average of quantity relatives weighted by reference to value (L4, P1) or to labour content (P33).

Q5   Quartiles
The three items which divide a statistical series arranged in order of magnitude into four equal parts. The first quartile is the 25th percentile (P11), the second, the median (M14), and the third, the 75th percentile. The term *second quartile* is never used, so that the common practice of referring to the first as the *lower quartile* and

the third as the *upper quartile* may be regarded as grammatically defensible. Half the difference between the upper and lower quartiles is the *quartile deviation,* which is one of the measures of dispersion (D19).

Q6   Questionnaire
A document containing a list of questions, copies of which are sent to a number of people or organizations who are invited to provide the answers, or required to do so under the provisions of a statutory instrument, such as the Statistics of Trade Act, 1947, or under the membership rules of a trade association or similar body. For a consideration of the problem of bias arising from a voluntary statistical questionnaire, see sampling (S3).

Q7   Queueing theory
Seeks to find means of solving problems of congestion and, in particular, of reducing congestion, which is regarded as taking the form of a queue. Machines awaiting repair in a factory are regarded as standing in a queue; so, too, are the orders kept waiting as a result of those machines being out of service. There are three basic statistical figures in most queuing situations: (1) the average number in the system, n; (2) the average rate of arrival, x; (3) the average rate of departure, which is equal to the average rate of service, y. There are two broad approaches to the problem: the analytical, which involves the use of mathematics; and simulation, the Monte Carlo method (see below).

With the analytical approach, the equations below may be required to arrrive at a solution. It is assumed that $y > x$, so that there is a tendency for the queue to diminish:

Average number in the system, ie in the queue and being serviced:

$$\frac{x/y}{1 - (x/y)} = \frac{x}{y - x}$$

Queueing time, in terms of the unit of time in rates of which x and y are expressed:

$$\text{Expected} = \frac{1}{y - x} - \frac{1}{y}$$

$$\text{Average} \quad = \frac{x}{y} - \frac{1}{y-x}$$

Queueing plus service time, average:

$$\frac{1}{y-x}$$

To reduce queueing time, the firm must either increase the average rate of departures or reduce the rate of arrivals. It is unlikely to have any control over the latter, and it may find it expensive to control the former. With these equations, the firm would be able to assess the effect of a given increase in the rate of departures. It must then set off the potential benefit of this effect against the expense, and reach a decision on this evidence. The potential benefit may be expressed in terms of annual gross revenue, and the expense in terms of annual cost.

Many queueing problems may be solved by the less mathematical, more statistical Monte Carlo method, which may in some cases be applied in tandem with the analytical method. The basic principles of Monte Carlo can best be demonstrated with a simple numerical example.

Suppose a cashier in a supermarket takes an average of three minutes to serve a customer. Customers arrive at the cash desk at a random rate of 15 an hour. How long does each customer wait in the queue on average? If the flow of customers is even, there would never be any queue. But, realistically, the flow is rarely even; customers arrive at the cash desk at random (R3). An idea of the randomness could be gleaned by inspection with the aid of a stop-watch, but this is a time-consuming exercise. The arrival of customers can be simulated by the use of random numbers, of which published tables are available. The hour is divided into a suitable number of equal parts. In this case the minute will suffice for demonstration purposes, though a tenth of a minute would probably be more appropriate. Of the 60 random numbers from 1 to 60, the first 15 are taken to represent the time in the hour of the 15 arrivals. If 2 is one of the numbers selected, then it is assumed that one customer arrived at the cash desk during the second minute of the hour. The 15 selected numbers are arranged in ascending order of magnitude, and entered in the first column of a table with four columns as in Table Q7.1. Where previous $F - A$ is negative, the entry under D is shown as zero. The entries in the columns headed A, S and F are all clock times, whereas those in the column headed D are periods of time in minutes. The average queueing time is the total of column D divided by 14, ie $15 - 1$.

Table Q7.1   The Monte Carlo method.

Arrival time	Delay	Service times	
	Previous $F - A$	Start	Finish
A	D	$S = A + D$	$F = S + 3$
02	0	02	05
05	0	05	08
06	2	08	11
10	1	11	14
11	3	14	17
16	1	17	20
. . .		. . .	

For the logic of the method, consider the succession of customers. The first arrives at the shop when it opens at 9 am, arrives at the cash desk at 9.02 with her purchases, is served immediately and leaves at 9.05, just as the second customer is arriving at the cash desk. She, too, is served immediately and leaves at 9.08. In the meantime, at 9.06, the third customer has arrived at the cash desk and she is kept waiting until 9.08, ie for 2 minutes, the entry in column D.

However, the method as demonstrated is not realistic. The actual service time may range from 1-10 minutes. For the purpose of making the analysis more realistic, the investigator could reasonably assume there is a Poisson distribution (P19), and use a second set of random numbers to indicate how the various service times succeed each other during the hour. A random succession may be obtained more easily by drawing from a hat. If the distribution shows that one service of one minute is provided in the hour, three services of two minutes, four of three, two of four, and so on, then there would be one slip of paper in the hat marked 1 (minute), three marked 2 (minutes), four marked 3, and so on; 15 slips of paper in all to be drawn from the hat to indicate the random succession.

Then there is another average to be examined: the average of 15 customers an hour. Both the average and the maximum number vary from hour to hour and day to day. Both may be very small on Mondays during the first hour, and at their peak on Saturday during the third hour. There is no doubt an element of random variation here but, for the most part, the distribution would conform fairly closely to a weekly pattern, which can be introduced into the analysis.

It has been stated that, to apply the Monte Carlo method satisfactorily, a computer is necessary. This may be true in some cases where repeated trials have to be made to obtain representative data for analysis, but not by any means in all.

Applications. Factory layout and reorganization; progressing documents in an office; organizing a transport and distribution system; timing traffic control lights.

Q8   Quick ratio

A business ratio (B17) whose current terms are readily available, usually current assets to current liabilities. Sometimes called the *acid ratio.*

Q9   Quota sampling

See public opinion polls (P38)

R1   Rag-bag
A facetious term for a residual heading in a written classification (C13), or a miscellaneous bin in a factory.

R2   Railway goods traffic receipts
The sales proceeds from goods transport services provided by the railways. They exclude parcels, etc. traffic by passenger train. Regarded at one time as an indicator of the trend of the country's prosperity. In the USA the figures used for the same purpose were freight-car loadings. Official production index numbers (I7) have taken their place.

R3   Randomness
Haphazardness; a state of numerical affairs that makes statistical methods worthwhile.

Random numbers are figures which bear no relation to each other except in so far as they fall between two extremes. They have a random order, though for some purposes a selection may be rearranged in order of magnitude. Used in simulation (Q7) and drawing random samples.

Random sampling. The drawing of a sample (S3) in such a way that every item in the universe (U3) is given an equal chance with all other items of being selected.

Random variable is an item whose magnitude cannot be precisely predicted owing to the effect on it of indeterminate or unknown factors. The time of sunrise at Greenwich is not a random variable, nor is a series of figures giving the time of sunrise on each day in

1973 a statistical series: their calculation falls within the sphere of mathematics, not statistics.

Random walk. A figure of speech denoting an unpredictably erratic progress from one state or magnitude to another.

R4   Ranking

Arranging a number of people or things in order of merit by reference to some other judgement criterion. Sometimes used in the sense of the more precise 'calculating rank correlation' which is concerned with the coefficient of correlation (C25) between two sets of rankings of the same thing or related things, of which the statistical series consist of rank numbers and not merit marks. The most-used formula, known as *Spearman's coefficient of rank correlation,* is

$$1 - \frac{6\Sigma(x-y)^2}{n(n^2-1)} = 1 - \frac{6\Sigma(x-y)^2}{n^3-n}$$

where Σ is a sign of summation; x represents one series of rank numbers, y the other series, paired so that each pair relates to the same entity; and n is the number of pairs. It will be seen that, where the two sets of rankings are identical, the coefficient is +1.00. Where the rankings are in complete disagreement, the coefficient is −1.00.

Subjective ranking judgements made by two judges in, for example, a beauty contest or a fashion show, or by people assessing colour appeal, design, packaging,· types of advertisement, security measures, etc. may be tested for agreement by using the coefficient of rank correlation.

Objective ranking calls for the use of a standard criterion of measurement. One criterion applied to some items under scrutiny, and other criteria to other items may give misleading results. Choice of criterion may sometimes be a matter of subjective judgement, but whatever the choice may be, it should be adopted as the standard. A failure to recognize this principle of objective ranking has caused much confusion of thought in at least one area of the management scene, that of investment appraisal. See yield criteria (Y3).

R5 Ratio delay-activity sampling

The ratio of the number of occasions on which a process, machine or plant is found to be idle to the total number of occasions on which observations are made. The observations are made at predetermined points of time at regular intervals during the working day or shift, and the ratio may apply to all factors causing idleness or delay, or to any particular factors such as the formation of a queue (Q7). If D is the ratio, d the number of idle occasions, and n the total number of observations, then

$$D = \frac{d}{n} \tag{5.1}$$

The standard error, S (S25), of the ratio is given as

$$S = \frac{d}{n^2}(1 - \frac{d}{n}) = \frac{D}{n}(1 - D) \tag{5.2}$$

which applies to large samples, at least of 100 observations.

Applications. In exercising the management control function; and for calculating contingency allowances for the time required to carry out a process.

R6 Rational price fixing

The fixing of prices that maximize profits. The following formula is used to derive the optimum price, as the rational price is often called:

$$p_m = a\frac{e}{e-1} \tag{6.1}$$

where p_m is the optimum price, a the marginal cost, and e the price elasticity of demand. Its derivation is given under marginal revenue (M6). See also I2, M4, N6, P27, S27.

Applications. Marketing and financial policy.

R7 Regression analysis

A statistical analysis designed to determine the extent to which one factor changes with changes in another factor or other factors. There are several methods of regression analysis, and some of the more useful ones are defined and demonstrated under the following entries:

C36 Cross-classification
F3 Finite differences
G5 Graphing
L7 Least squares
T16 Trend ratios

See also L13, R8 and 9.

Applications. Calculating marginal cost, elasticity of demand, learning and scale of production laws, and similar values.

R8   Regression coefficient
A coefficient in a regression equation (R9).

R9   Regression equation
The solution equation of a regression analysis (R7).

R10   Relative measure
An average or ratio, as distinct from absolute measure (A2). Most derived statistics (D11), including all index numbers (I7), are relative measures.

R11   Replacement cost
The current cost of materials and bought-out parts and components; the expected or forecast cost of replacing a wasting fixed asset on life expiry. Both have an important part to play in the calculation of company net profits, especially in periods of inflation and deflation. In this latter respect, replacement-cost accounting has a serious rival in a technique proposed by the Accounting Standards Steering Committee in its Exposure draft No. 8, published by – among others – the Association of Certified Accountants, as a technical supplement to its monthly journal the *Certified Accountant* (February 1973). In this, the currently weighted (Paasche, P1) annual consumers' price index is used for adjusting figures on the conventional historical basis to a 'current general purchasing power basis'. Management statisticians may find much of practical interest in the ASSC's proposals.

Inflation or no, the replacement cost of fixed assets has a part to play in cost accounting. There are two concepts: (1) the gross replacement cost; (2) the net replacement cost. Oddly enough, the latter is sometimes greater than the former. The gross replacement cost of an asset is the forecast total cost of providing a replacement asset, like for like, when the old asset reaches the end of its economic life. The net replacement cost is the forecast gross replacement cost plus the forecast cost of dismantling and removing the old asset, less the forecast residual value of the old asset. Where the cost of dismantling and removal exceeds the residual value, as it almost invariably does with buildings, the net replacement cost exceeds the gross replacement cost. It is the net replacement cost which investment appraisers and cost accountants regard as the total amount to be provided under their annual depreciation charges. For premature displacement cost, see P22, U4; for sinking-fund depreciation, see S18.

Applications. Accounting for changes in the value of money, investment appraisal, cost accounting.

R12　Reporting

Presenting to management the results of an experiment, a piece of research, or the state of affairs in a department, a section, or the company as a whole. Even where the statistician is to report to a committee, he may find it advisable to prepare a written report for circulation prior to, or at the meeting of the committee. A written report serves as a permanent record; it can be read and studied at any time convenient to the recipient, and it provides a means of communication that in many respects cannot be bettered. A good written report has the following attributes:

1　It is readable and interesting.
2　It presents the facts and conclusions clearly and concisely.
3　The facts are stated objectively and arranged in an order that best meets the requirements of the inquiry and the argument; all opinions are stated as such, and the people who have expressed them are named.
4　The conclusions are logical inferences drawn from the facts as stated. Where they are numerical, the standard error (S25) is given where appropriate.
5　Jargon is avoided, and technical terms are used sparingly and only when it is known that the reader will understand them.

6   The opening paragraphs identify the object of the inquiry, the terms of reference, and the authority, if any; the middle paragraphs give the relevant, and only the relevant facts, as ascertained and verified, and the concluding paragraphs discuss the facts and state the conclusion and recommendations.

In long reports, it is often a convenience to busy readers to have the conclusions and recommendations stated in an early paragraph, preferably the one immediately following the identification paragraphs.

There are many literary pitfalls and logical lapses for the unwary. Some of the latter are touched on elsewhere: under C7, D4, I8, N7 and 8, S28 and 40. The former are less a matter of grammatical purity than of clarity and conciseness. It is more important to avoid ambiguity and circumlocution than it is to avoid a split infinitive: though, to be sure, splitting infinitives is not a practice to be recommended.

Writers on the subject give more importance to style than it appears to merit. A detached, self-effacing style seems to be the one worth striving for, and probably the best way to achieving it is to 'play the sedulous ape' to the *Times* Law Reports, as R. L. Stevenson ('A college magazine', in *Memories and Portraits*) did to Hazlitt, Lamb, Defoe and others. For notes on the presentation of statistics in reports, see D15, P10, S32, T1 and 2.

R13   Representativeness
The extent to which the average (A16) and the standard deviation (S24) of a sample (S3) approximate to the average and standard deviation of the universe (U3) from which the sample is drawn. See standard error (S25).

R14   Residual heading
A heading in a statistical classification (C13) designed to render the classification, or a section of it, exhaustive. It is usually worded 'Other', or 'Other types', or 'Other kinds of XYZ', or 'Other sizes of XYZ' or 'XYZ not elsewhere specified'.

R15   Retail-price index
Formerly the cost-of-living index (C30); compiled monthly and

published monthly through the press and in full detail in the *Department of Employment Gazette.* Not to be confused with the consumer price index, which is an annual currently weighted index (Paasche, P1), compiled by the Central Statistical Office specially for correcting national income statistics for changes in the purchasing power of money (P39).

R16   Revenue method

An alternative to the capital method of accounting (C3) for expenditure and receipts of a capital nature accruing after a new capital work has been brought into use. For appraisal purposes, such expenditure and receipts are regarded as current-account items under the revenue method.

Applications.   Investment appraisal, efficiency auditing.

R17   Revenue productivity index

A productivity index (P33) that rejects value added as the measure of output, and adopts instead a production index (I7) based on the value of production index divided by the price index of the company's output weighted by reference to values in the base period. This would give a currently weighted production index (Paasche, P1).

There are additional complexities, the object of which is not clear, designed to take care of all factors contributing to changes in productivity. Changes in the productivity of labour may take account of all factors without violating the principles of logic: an index showing an increase in the average output per man hour or per man year is a statistical measure; that it is expressed in such a way does not mean that it ascribes the increase to the labour force — that the labour force is working that much harder.

For the rest, it is proposed here to show that a value-weighted production index of a company's products probably does not provide so good a measure of physical output needed for making a productivity index, as does a labour-weighted production index, as described under productivity (P33). Table R17.1 contains a numerical example of a firm with two products, A and B. The notation is that commonly used in the literature, with the addition of *m,* which represents the labour content per unit of output. It will be seen that $m_n = 0.667m_0$ for both products, so that it can be said without further ado that the correct productivity index for period n

Table R17.1   Calculating productivity index numbers.

Factor	Product		
	A	B	Total
q_0	8	2	—
p_0	3	10	—
m_0	3	6	—
p_0q_0	24	20	44
m_0q_0	24	12	36
q_n	12	4	—
p_n	6	10	—
m_n	2	4	—
p_nq_n	72	40	112
m_nq_n	24	16	40
p_0q_n	36	40	76
p_nq_0	48	20	68
m_0q_n	36	24	60
m_nq_0	16	8	24

Production index numbers $(Q_{n.0})$

A   Laspeyres's value-weighted: $\dfrac{\Sigma (p_0q_n)}{\Sigma(p_0q_0)}$ ie $\dfrac{76}{44}$ = 1.727

B   Paasche's value-weighted: $\dfrac{\Sigma(p_nq_n)}{\Sigma(p_nq_0)}$ ie $\dfrac{112}{68}$ = 1.647

C   Fisher's ideal (GM of Laspeyres and Paasche) 1.687

D   Laspeyres's Labour-weighted: $\dfrac{\Sigma(m_0q_n)}{\Sigma(m_0q_0)}$ ie $\dfrac{60}{36}$ = 1.667

E   Paasche's labour-weighted: $\dfrac{\Sigma(m_nq_n)}{\Sigma(m_nq_0)}$ ie $\dfrac{40}{24}$ = 1.667

Productivity index numbers

$Q_{n.0}$ $\dfrac{\Sigma(m_0q_0)}{\Sigma(m_nq_n)}$ ie $\dfrac{36}{40}$ = 0.9

A   0.9 of 1.727 = 1.554
B   0.9 of 1.647 = 1.482
C   0.9 of 1.687 = 1.518
D and E   0.9 of 1.667 = 1.500

on period 0 (= 1.00) is 1/0.667, ie 1.500. Its calculation by the standard methods are shown in the table. Both Laspeyres's (L4) and Paasche's value-weighted index numbers of production are included, and Fisher's ideal (F4) also. It will be seen that only the labour-weighted production index provides the appropriate numerator of what is known to be the correct productivity index. It therefore seems that the revenue productivity index is not to be recommended. In common with some other elaborately complex methods, it seeks a measure which in the last resort is profit, the determination of which is a function of the profit and loss account, not of a statistically contrived productivity index.

The entry on substitution (S38) throws some light on the direction of the relative bias in a comparison of Laspeyres's and Paasche's value-weighted index numbers.

R18   Rounding error

In industrial statistical and costing offices, it is a common practice to derive averages for general use. The cost per vehicle mile of tyres and of lubricants of a given type of road motor, and the mpg of fuel are examples. There are two errors in nearly all average of series: (1) the rounding error; (2) the standard deviation (S24) or the standard error (S25), the latter for a probability of 68%, or twice the standard deviation or error for a probability of 95%.

The rounding error can be reduced by calculating the average to more significant figures. Where an average is given to two significant figures, eg 27 000, it can be regarded as meaning 26 500-27 500, whereas if it is given to three significant figures, say, 27 300, it means 27 250-27 350. In the former the rounding error is 500, whereas in the latter it is 50, both with 100% probability. As there is no reason to suppose that the true average lies nearer to the central average, as calculated, than to either extreme, or to any other figure between the two extremes, it seems that one cannot logically assume that the rounding error conforms to the normal curve (N9), and is therefore equal, near enough, to three standard deviations; it is thus impossible to calculate the error for one standard deviation with a probability of 68%.

There may not be much point in extending the average of a series to more than three significant figures where the standard deviation of the series is equal to 0.10 or more of the average itself. See coefficient of variation (C14). The rounding error would be of little significance in such circumstances. If the average is given as 27.2, the rounding error would be 0.05 compared with a standard deviation of 2.72 or more. However, it is better to extend an average to more, rather than fewer significant figures than are justified.

Not all averages have a standard deviation. Ratios of basics or firm figures such as the P/E ratio (P9), the dividend yield (D21), and the times covered (T7) have a rounding error but not a standard deviation, and they can therefore be extended to any number of significant figures.

The application of rounded averages to practical purposes calls for a measure of care. Probably the best approach is to calculate the required figure based on each of the two extremes, as well as the averages as given. Multiplication of rounded averages or ratios may increase the error very appreciably. The calculation of the product of two averages may be set out as follows

	Averages	Extreme values		Error
		Lower	Upper	(%)
	23	22.5	23.5	2.17
	45	44.5	45.5	1.11
Product	1 035	1 001.25	1 069.25	
Error of product		−33.75	+34.25	3.28

The mean rounding error of the product is equal to half the sum of the averages. Where one figure is a basic and the other an average, the rounding error is equal to the product of the basic and the absolute rounding error of the average.

S1 Sales analysis
The analysis of market demand (see D7); for method of analysis, see R7.

S2 Sales schedule
Demand schedule (D8).

S3 Sampling
Determining the properties of a universe (U3) by examining or testing a small part of it. The accuracy of the information thus obtained depends on the size of the sample, the way it is drawn or selected, and the homogeneity of the universe. Ideally, a sample should be relatively large, its selection should be random, and the universe from which it is drawn should be complete and homogeneous. Random selection from a complete universe gives every item in the universe an equal chance of selection.

Whether a sample is large enough can be judged subjectively by reference to the sampling error (S4), which gives an estimate of the standard deviation (S24) of the means of samples of the same size drawn from the same universe. For any given standard deviation of the sample, the sampling error is smaller for large samples than it is for small samples.

The objective of sampling in general is to obtain reasonably accurate data at a much lower cost than that of inspecting or testing every item in the universe. The objective of carrying out a particular sampling inquiry must be clearly and precisely stated or at least understood.

Most problems encountered in practical sampling are considered elsewhere in this work under the following headings:

One matter remains for consideration here: sampling by postal questionnaire. No matter how carefully and randomly samples are chosen, postal surveys are biased beyond measure. Responders to a statistical questionnaire, and non-responders as groups, each have their own peculiarities, and some of these peculiarities may be relevant to the objects of the survey. Nobody knows these peculiarities, and it is also probably true to say that responders rarely account for as much as 50% of the sample. It is not enough to double or treble the size of the sample to make good the non-responders: that will not eliminate the bias.

Applications. Market research, quality control, purchasing policy.

S4   Sampling error

An estimate of the standard deviation of the averages of a number of samples of the same size drawn from the same universe, based on the data of one sample. It is equal to the standard deviation (S24) of the sample, divided by the square root of the number of items in the sample less one:

$$S = \frac{\sigma}{\sqrt{(n\text{-}1)}} \tag{4.1}$$

The formula applies to a 68% probability, and doubling the sampling error gives a 95% probability that the average of any sample of the same size will fall within twice the sampling error of the average. The formula assumes that the sample is not biased. See normal curve of error (N9); as applied to index numbers, see I7; also see N11.

Applications. Random sampling, including quality-control sampling.

S5   Sampling fraction
The actual or estimated ratio of the number of items in a sample to the total in the universe. A 5% sample has a sampling fraction of 1/20. See grossing up (G8).

S6   Sampling universe
See universe (U3).

S7   Scale of production law
See percentage laws (P10).

S8   Scatter
A visual indication of the correlation existing between two series of statistics. The extent to which the dots in a scatter diagram (S9) fail to fall into alignment.

S9   Scatter diagram
A graph showing the scatter of plotted dots, as in Figures B5.1 and F3.1.

S10   Seasonal chart
A chart showing plotted time series, month by month or quarter by quarter, for each of a number of years. Figure S10.1 contains a seasonal chart of quarterly figures for three years. To complete the seasonal pattern, the figure for the last period of each year is carried forward to form the beginning of the following year.

Applications. A preliminary to estimating regular seasonal variations (S11).

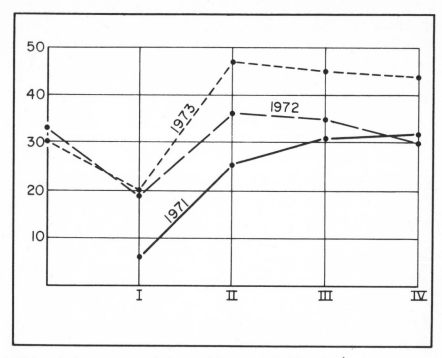

Figure S10.1   Seasonal chart of figures in Table S11.1.

S11   Seasonal variations

Variations in sales turnover, current costs, etc. due to regularly recurring seasonal factors. The standard method given in most elementary textbooks of statistics for determining seasonal variations is good, except where there is evidence of a general upward or downward trend. Where there is a trend in the basic statistics, then unless the trend is first eliminated, the calculated seasonal variations will be affected by the trend. Table S11.1 shows how a linear trend can be eliminated, and the seasonal variations calculated. Generally, monthly data are used but quarterly data are adequate to demonstrate the method.

It will be seen that the trend norm swings round the overall average of 30, since its own average is also equal to 30. The successive differences in the norm are all 2.250, which is derived from the averages of the three years:

$$\frac{30-21}{4} = \frac{39-30}{4} = 2.250$$

Table S11.1   Seasonal variations and trend.

Quarters	1971	1972	1973	Average	Trend norm	Seasonal variations	
						$(5) - (4)$	% of (4)
	(1)	(2)	(3)	(4)	(5)	(6)	(7)
I	6	19	20	15	26.625	+11.625	+77.5
II	25	36	47	36	28.875	−7.125	−19.8
III	31	35	45	37	31.125	−5.875	−15.9
IV	32	30	44	32	33.375	+1.357	+4.3
	84	120	156	120	120.000	0.000	
Quarterly average	21	30	39	30	30.000		

which also indicates that the trend is linear. The sum of the trend norms for quarters II and III is 60, ie, twice the average. The figures of 28.875 and 31.125 are calculated by simultaneous equations. Let quarter III be represented by x, and quarter II by y, then

$$x + y = 60.00$$
$$x - y = 2.25$$

Therefore $\qquad 2x = 62.25$

Therefore $\qquad x = 31.125$

and $\qquad y = 28.875$

Since the seasonal variations are likely to increase in absolute terms as the upward trend continues, it is better to use the percentage figures in the last column for seasonal correction than the figures in column 6 of the table. If the basic figure for quarter I of 1974, or of any future year, turns out to be 30, then the seasonally corrected figure would be

$$30 + 77.5\% = 53.25$$

Applications.   Making seasonal corrections to data such as those used in calculating business ratios, and in budgeting.

S12   Sectors of the economy

A classification of the economy used in the national accounts. There are six main sectors:

1   Public sector, comprising central and local government and public corporations including the Issue Department of the Bank of England, but not the Banking Department.

2   Banking sector, comprising the UK offices of deposit banks, including the Banking Department of the Bank of England, accepting and discount houses, overseas banks and the National Giro.
3   Other financial institutions comprising hire-purchase companies, building societies, trustee savings banks, the Post Office Savings Bank, insurance companies, superannuation funds, unit and investment trusts, etc.
4   Companies sector, comprising privately controlled profit-seeking corporate enterprises, excluding banks and other financial institutions.
5   Personal sector, comprising non-profit-making bodies, private trusts and unincorporated enterprises. Life assurance and pension funds are regarded as assets of this sector.
6   Overseas sector, defined in the *National Income* blue book in terms of the transactions between UK authorities and private residents, on the one hand, and overseas authorities, etc., on the other, on both capital and current accounts. Sectors 2, 3, 4 and 5 are sometimes combined under the heading 'Private sector'.

S13   Secular trend
Long-term trend; trend over a number of years. See S39.

S14   Sequential testing
A technique of quality control (Q3) used where testing is expensive, eg where it involves destruction in testing for length of life. A minimum sample is drawn from a batch and tested. If all items in the sample pass the test the batch is accepted; if not, further smaller samples are drawn and tested until it can be finally decided that the batch may be accepted or must be rejected.

S15   Semivariable costs
Costs that are partly variable and partly fixed. Repair costs are in many cases semivariable. See fixed costs (F8), marginal cost (M4) and variable costing (V5).
   Applications.   Product costing, cost analysis.

S16   Sensitivity analysis
The determination of the risk factors involved in a management decision, ranking them in order of importance, taking the most

optimistic and pessimistic forecasts of the consequences of the decision, and attempting to calculate the probability that the consequences will fall midway between the two extremes.

Applications. Management decision making.

S17 Simulation

The operation of a model, mathematical or physical, designed to represent a real, or potentially real situation, with a view to determining the consequences or to training. In statistics, risk and randomness are usually introduced by using random numbers. An example is the Monte Carlo method (Q7). Another example is the game 'Monopoly', which simulates the property market, with the throwing of dice used for introducing an appropriate element of chance and risk.

Applications. Management decision making, management development.

S18 Sinking fund

A fund actually or notionally accumulating at compound interest into which annual payments of equal size are paid; set up to provide a given sum for redeeming a dept maturing at some date in the future, or for replacing a wasting fixed asset when it falls due for renewal, or for amortizing the outlay on an asset that may not be renewed, such as a leasehold property.

The *annual sinking fund* is the annual amount paid into a sinking fund. Published tables are available showing the annual sinking fund required to provide £ at the end of n years at $R\%$ interest. The formula is

$$\text{S F to provide £ in } n \text{ years} = \frac{r}{(1+r)^n - 1}$$

where r is the rate of interest per £. It will be seen from A9 that this formula is the reciprocal of the amount of £ a year. It also forms part of the present value of £ a year (P23) which is equal to the reciprocal of the annual sinking fund plus the rate of interest per £.

It has been said that the sinking fund is a cost-accounting fiction, but a convenient one. There is truth in that, yet it is a wise sole proprietor who sets up a real sinking fund to provide finance for the replacement of his cab, his leather-stitching machine or his printing press, when it falls due for renewal. Relevant entries are as follows :

A8	Amount of £
A9	Amount of £ a year
A11	Annual capital charge
A12	Annual value
A13	Annuity
C16	Compounding
D10	Depreciation
D16	Discounting
I14	Internal rate of return
I17	Investment criteria
N4	Net present value
P23	Present value
R11	Replacement of fixed assets
U4	Unprovided renewals

Applications. Investment appraisal, efficiency auditing, pension-fu calculations.

S19   Size of firm

Size of subsidiary company, parent company or holding company in terms of capital employed (C2), sales turnover, net assets, labour force, or factory floor area, whichever is relevant or merely convenient.

S20   Skewed distribution

An asymmetrical frequency distribution (F9). Skewness has a number of measures:

Absolute measures

$$\text{Average minus mode}$$

$$\text{Average minus median}$$

$$\sqrt[3]{\frac{(\text{Sum of mean deviations})^3}{\text{Number of items}}}$$

which is called the third moment (M18).

Relative measures (coefficients of skewness)

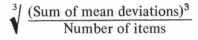

$$\frac{\text{Average minus mode}}{\text{Average deviation}}$$

$$\frac{\text{Average minus mode}}{\text{Standard deviation}}$$

$$\frac{\text{Average minus median}}{\text{Average deviation}}$$

$$\frac{\text{Average minus median}}{\text{Standard deviation}}$$

$$\frac{\text{Third moment (as above)}}{\text{Standard deviation}}$$

Where the mode and median fall short of the average, the answers in all measures are positive, and the frequency distribution is then said to be positively skewed. Where they exceed the average the answers for all measures are negative, and the distribution is then said to be negatively skewed. Relevant entries are median (M14), mode (M15), average deviation (A17), standard deviation (S24), mean deviations (D19 and 13).

S21   Spot sample
A small sample taken on the spot without regard to its randomness or representativeness (S3).

S22   Spurious correlation
The introduction into a correlation (C25) or regression (R7) analysis of a theoretically irrelevant factor, or the introduction of two or more factors as independent variables (I6) which, in effect, provide measures of the same force.

Applications.   Analysis of time series.

S23   Standard costing
Analysing the variance (A10) of costs against standard costs, which are based on the efficient operation of the undertaking; also the preparation of standard costs.

S24 Standard deviation

The principal measure of dispersion (D19); generally symbolized by σ. The formula is

$$\sigma = \sqrt{\frac{\Sigma(x^2)}{n}} \qquad (24.1)$$

where x is the deviation of X from the mean value of X, n the total number of observations in the series, and Σ the sign of summation. It will be seen from the entries under normal curve of error (N9) and quality control (Q3), among others, that the standard deviation plays a central role in statistical theory and methods. For small samples, Bessel's correction (B4) is applied to give an estimate of the standard deviation in the universe (U3):

$$\sigma_u = \sqrt{\frac{\Sigma(x^2)}{n-1}} \qquad (24.2)$$

S25 Standard error

The measure of error in a parameter (P3) for a 68% probability. Twice the standard error gives the fiducial limits (F2) for a 95% probability. See normal curve of error (N9) for the full range of probabilities for various fractions and multiples of the standard error.

Formulae are generally specific to the type of parameter. For the average of a series or of a sample, the standard error is often referred to as the *sampling error*, the formula being

$$S_x = \frac{\sigma_x}{\sqrt{(n-1)}} \qquad (25.1)$$

where S_x is the standard error of the average of X, σ_x the standard deviation (S24) of the series of X, and n the number of observations. The greater the value of n, the smaller is the value of S.

For estimates of the dependent variable (D9) calculated from a least-squares (L7) regression equation (R9), the standard error is usually referred to as the *standard error of estimate*, the formula being

$$S_z = \sqrt{\left(\frac{\Sigma(z^2)}{(n-m)}\right)} \qquad (25.2)$$

where z is the difference between the actual value and the corresponding estimated or calculated value of the dependent variable, n the number of sets of observations, and m the number of variables in the analysis, $(n-m)$ being the degrees of freedom (D6).

The standard error of estimate is used in calculating the standard error of regression coefficients (R8) derived by least squares (L7). Strictly, neither the standard error of estimate nor the standard error of a regression coefficient applies where there is a significant degree of intercorrelation between the adopted dependent variable and any one or more of the independent variables. Both should be limited to cases where none of the independent variables is influenced by the dependent variable.

For the standard error of a regression coefficient, S_r, the formula is as follows:

$$S_r = S_z \sqrt{c} \qquad (25.3)$$

where S_z is the error of estimate — (25.2) above — and c is specially derived for the purpose. Subscripts are added to indicate the coefficient to which it refers, except in simple regressions when

$$c = \frac{1}{\Sigma(x^2)} \qquad (25.4)$$

in which x represents the deviations of X from the average of X. In multiple regression, c_{22} applies to the coefficient a_2, of X_2; c_{33} to the coefficient a_3, of X_3, and so on. The formula for deriving the value of c is that of Table L7.1 with the nth equation omitted and with 1 and 0 providing the values of the righthand side of equations (1) to $(n-1)$. For deriving the standard error of a_2, 1 is the righthand side of equation (1), and 0 in the righthand side of the other equations, and the equations are solved simultaneously to determine the value of c_{22}; for deriving the standard error of a_3, 1 is used in the righthand side of equation (2) and 0 in equation (1), and in the other relevant equations.

A problem in three variables would need two equations for simultaneous solution in each case. For evaluating c_{22}, they would be

$$\Sigma(x_2{}^2)c_{22} + \Sigma(x_2x_3)c_{33} = 1 \qquad (25.5)$$
$$\Sigma(x_2x_3)c_{22} + \Sigma(x_3{}^2)x_{33} = 0$$

and for evaluating c_{33}, they would be

$$\Sigma(x_2{}^2)c_{22} + \Sigma(x_2x_3)c_{33} = 0 \qquad (25.6)$$
$$\Sigma(x_2x_3)c_{22} + \Sigma(x_3{}^2)c_{33} = 1$$

where x represents the deviations of X from the average of X, x_1 applies to the dependent variable, and x_2 and x_3 to the two independent variables.

It is customary to present the standard error of a regression coefficient in parentheses immediately beneath the coefficient itself in the regression equation, and to give the standard error of estimate either in the text or beneath the unattached constant in the regression equation.

In solving for the values of c, it is important to see that the simultaneous equations are arranged in the correct order. A clue to which equation takes 1 on the righthand side is that the coefficient of c whose value is being sought is always the sum of the squares of the mean deviations of the corresponding independent variable in that equation.

The formula of the standard error of a ratio delay-activity sample is given in R5.

Applications. In calculating upper and lower limits, for any desired probability, of parameters, index numbers, and, therefore, of entities such as the optimum price.

S26   Standard official classifications

Classifications of general use in compiling official statistics. The classification of occupations is specific to the censuses of population. There is no official commodity classification of the United Kingdom; each kind of statistics having its own classification, namely

Products Census of production (C9)
Exports Export List (E10)
Imports Import List (I3)

The census of production commodity classification follows the lines of the Export List, though not exactly. It is understood that some attempt has been made to reconcile the three commodity classifications, presumably with a view to compiling a standard classification of commodities.

The *Standard Industrial Classification* (SIC) is a classification of industries ranging from agriculture, fishing and forestry (Order I) to public administration and defence (Order XXVII). The 27 orders are divided into 181 classes — *Minimum List Headings*. The classification is used for labour statistics and for the censuses of production and distribution. It was first published in 1948 by HMSO, was revised in 1958, and again in 1968. Establishments are classified against the minimum list headings by reference to their major products, as returned in their latest census forms. HMSO publishes an alphabetical list of industries defined in terms of products: it lists each product and shows the SIC minimum list heading to which it belongs.

S27   State industries financial policy

The policy of the state industries in respect of price fixing, the payment of interest on issued loan capital, debt redemption, and the provision of finance for the renewal of fixed assets and for capital growth. Government policy in recent years in respect of the state industries has tended towards the financing of debt redemption and capital growth, as well as the interest on debt and the renewal of assests, out of current revenue, much as the more financially prudent commercial undertakings try to do. The annual provisions for debt redemption and future growth, as well as interest and depreciation, can be regarded as annual costs forming part of the undertaking's fixed costs.

Figure B14.2 shows that there are two break-even prices, equal to the total sales proceeds divided by the total costs, and corresponding to the two points of intersection of the sales curve and the cost curve labelled B_1 and B_2. For B_1 the price is higher than for B_2, so that, in order to break even and to give consumers the best service at the lowest price, a state-owned non-profit-making undertaking would choose the lower of the two break-even prices, that corresponding to B_2. But if the annual fixed costs are loaded with provisions for debt redemption and growth, the cost curve, T will rise bodily to a new position, and the intersection B_2 will move along the sales curve to the left, to a point corresponding to a higher break-even price. If the cost curve were to rise above the sales curve on all ordinates, price increases or decreases, above or below the optimum, would tend to increase the annual loss rather than reduce it. Here the optimum price is defined as that price which minimizes the annual net loss.

S28   Statistical inference

A somewhat pretentious term when used in the sense of drawing numerical conclusions from premises based solely on statistical data. A statistical inference is just as much the fruit of general reasoning as any other kind of inference, and the same rules apply. For instance, a precise inference drawn from imprecise premises is a *non-sequitur* (N8), whether the inference and the premises are statistical or not. An example of a statistical *non-sequitur* is given under N8. Properly, a numerical conclusion drawn from a sample or census.

S29   Statistical notation

See notation (N10).

S30   Statistical quality control

See quality control (Q3).

S31   Statistical progression

A time series (T8) with a seasonal pattern (S11) or a secular trend (S13), or both. A *progressive average* is the AM (A14) based on a cumulative total; it is not a moving average (M21).

S32   Statistical schedule

A device which sets out the relevant figures of a dependent variable (D9) for ranges of the independent variables (I6); often based on a regression equation (R9). Examples will be found under D8, N6 and P17.

Applications. Presentation, analysis, extrapolation, including forecasting.

S33   Statistical table

An orderly presentation in tabular form of sets of statistical data. For classification, see C13, also see T1.

S34   Stochastic equation

An equation from which some independent variables (I6) are necessarily omitted because they are either unknown or have no numerical measure. The subject is discussed in C25.

S35   Stock control

See EOQ (E6).

S36   Stratified sampling
A sampling device by which the universe (U3) is divided into strata by reference to the size of a relevant and known factor, larger proportions of the more important stratum being drawn for sampling purposes than those of the less important. Each stratum has its own sampling fraction (S5), its own standard deviation (S24), and its own sampling error (S4 and 25). The object of stratified sampling is to minimize the size of the whole sample for any given level of representativeness (R13).

S37   Student's *t*
A ratio which is compared with the corresponding predetermined figure for given degrees of freedom (D6) and a given probability (P28) to indicate the representativeness (R13) of a sample, a regression coefficient (R8) or a regression equation. The pre-determined figure will be found in published statistical tables under the heading 'Student's *t*-distribution' or, merely, '*t*-distribution'.

For testing the significance of the average and standard deviation of a sample, the value of *t* is equal to the difference between the universe average and the sample average, sign ignored, multiplied by the root of the number of items in the sample less one, and divided by the standard deviation of the sample, ie

$$t = \frac{(\bar{X}_u - \bar{X}_s)\sqrt{(n-1)}}{\sigma_s} \tag{37.1}$$

where subscripts *u* and *s* indicate the universe and the sample, and the bars denote averages.

For a regression coefficient, the formula is

$$t = \frac{\text{Regression coefficient}}{\text{Standard error of the coefficient}} \tag{37.2}$$

The higher the calculated value of *t*, the lower the probability that it arises from chance, and therefore the more significant is the average, the standard deviation, or the regression coefficient.

S38   Substitution
The exploitation by purchasers, of relative changes in the prices of closely related goods and services with the object of maximizing

savings or minimizing losses. Consider a housewife who purchases 100 lb of butter and 20 lb of margarine a year when the prices are 20p and 16p per lb respectively. When the price of butter rises to 30p, she substitutes 80 lb margarine for butter. We then have the position shown in Table S38.1, where p and q represent the price and quantity purchased, respectively, and the subscripts 0 and n indicate the base period and given period. For a discussion of Laspeyres's and Paasche's index numbers, see L4 and P1, and for a list of related articles, see I7. It will be seen that Laspeyres's index number has an upward bias relative to Paasche's. Where substitution takes place owing to relative price changes, this apparent upward bias always exists.

It will also be seen that despite the rise in the price index of the two commodities combined, the housewife has contrived by substitution to reduce her expenditure on them from £23.20 a year to £22.00.

Applications. Sales forecasting and planning.

Table S38.1 Effect of substitution.

	Butter	Margarine	Total (p)
p_0	20	16	
q_0	100	20	
$p_0 q_0$	2 000	320	2 320
p_n	30	16	
q_n	20	100	
$p_n q_0$	3 000	320	3 320
$p_n q_n$	600	1 600	2 200
$p_0 q_n$	400	1 600	2 000

Laspeyres's price index:
$$P_{n.0} = \frac{\Sigma(p_n q_0)}{\Sigma(p_0 q_0)} = \frac{3\,320}{2\,320} = 1.43$$

Paasche's price index:
$$P_{n.0} = \frac{\Sigma(p_n q_n)}{\Sigma(p_0 q_n)} = \frac{2\,200}{2\,000} = 1.10$$

Fisher's ideal (F4)
$$P_{n.0} = \qquad\qquad 1.25$$

S39   Sunspots and the trade cycle

During the nineteenth century both sunspots and world trade had cycles of 11 to 12 years. With his propensity to argue *post hoc ergo propter hoc* (see C7, N7), man attributed the latter to the former. The protagonists of the theory contended that sunspot activity affected harvests, and it was through this that it caused the trade cycle.

Sunspot activity has had a cycle of about 10 years since the turn of the century, but the trade cycle has all but disappeared since the end of World War I; what little evidence there is suggests a cycle of about five years. None of this is to say that there is no causal relationship between sunspot activity and the trade cycle: there is no definite proof either way. Table S39.1 of sunspot activity is given for the information of those sales forecasters who think there may be something to be said for the idea. From it they may venture, without much risk, to forecast the years of maxima and minima for a decade or two, see C40, S13.

Table S39.1   Sunspot cycles since 1900.

Years of maximum activity	Years of minimum activity
1907	1901
1917	1913
1928	1923
1937	1933
1947	1944
1957	1954
1968	1964

S40   Syllogism

A common kind of deductive reasoning (D4), which, in its complete form, consists of two premises and a conclusion, and comprises three terms, eg

All men are mortal.
All kings are men.
Therefore all kings are mortal.

One term, *men*, appears in both of the premises, and is called the middle term. The term *mortal* is called the major term, and it appears in the first or major premise and the conclusion. The term *kings* is called the minor term: it appears in the minor premise, and forms the subject of the conclusion. It will be seen that if the two premises are true, the conclusion is also true (N8).

S41   Systems analysis

The examination, integration and rationalization of office systems and internal routine written communications prior to transferring the work to an electronic computer. Systems analysis alone has often resulted in appreciable reductions in annual costs, so much so that it is now recognized as a management technique in its own right without benefit of EDP. See I9.

T1   Table twisting
A tabulating process designed to present figures in the most effective way, either as information, or in furtherance of an argument. The process consists of twisting the table so that the stub (ie the line headings) becomes the column headings, and vice versa. See S33, T2.
   Applications.   Statistical presentation, report writing.

T2   Tabular presentation
The systematic presentation of statistics (S33, T1).

T3   Terminal valuation
An alternative to present value (N4, P23) of making a lump-sum evaluation of a capital asset. It compounds the capital outlay, the annual costs and the revenues to the year in which the asset is expected to be displaced. The formulae employed are those of the amount of £ and the amount of £ a year (A8 and 9) as necessary. For other methods of evaluating a capital asset, see A12, I14, N4, T15; also see C16 and D16. The concept seems to be generally regarded as illogical
   Applications.   Investment appraisal; ranking mutually exclusive alternative capital assets.

T4   Terms of trade
The ratio of the price index of exports to that of imports. A rise indicates a favourable change.

T5   Tests of significance
See chi-squared test, *F* test and Student's *t* (C12, F10, S37).

T6   Time chart
In statistics, a diagram showing the plotted points of a time series, measured on the vertical scale, with time on the horizontal scale. A seasonal chart (S10) is a form of time chart.

Applications.   Presentation, also useful for estimating time lags.

Γ7   Times covered
Sometimes called simply *the cover* or the *E/D ratio:* the ratio of the total earnings of a company to the total amount distributed as dividends, which is the same as the earnings per share to the dividend per share. See D21, E1, P9.

Applications.   Investment analysis.

T8   Time-series analysis
Not a distinct, single technique, but a set of techniques, the choice of which depends on the periodicity of the series, the number of observations and the purpose of the analysis. The techniques include analysis by moving averages (M21), by statistical progression (S31), by seasonal variation (S11), and by fitting a trend line (G6, L7 and L13, T16 and T17)

T9   Time trend
The trend of a time series (T8).

T10   Ton mile
A statistical unit of freight transport; similar to the unit of passenger transport, the passenger mile (P7).

Application.   Distribution cost analysis.

T11   Trade cycle
See S39.

T12   Tranche
A share, literally a slice.

T13   Transfer pricing
An internal pricing system for goods supplied and services rendered by one department of a company to another department, or from one member of a group of companies to another member. It is a

controversial subject. One school of thought argues that a transfer price should be a fair market price; another that it should be equal to the marginal cost (M4) to the supplier. If it is the company's policy that each division should be a profit centre and allowed to maximize its own profits without regard to the profit of the company, then undoubtedly the fair market price is correct, or at any rate, it conforms to the company's policy. But if the more logical policy of maximizing the profits of the company as a whole is pursued, then it seems that the correct transfer price is equal to the marginal cost. The maximization of central profits does not follow from the maximization of the profits of the individual divisions, since one division may increase its profits at the expense of a greater decrease in the profits of another.

Transfer pricing is tied up with the external pricing system. If the company pursues a rational pricing policy (R6), but allows the divisions to calculate their own rational selling prices, then if a purchasing department is to maximize the central profits of the company, it must base its selling prices on the company's marginal costs, which are those of the production divisions.

Where each division is a profit centre, there is the problem that either of the two divisions concerned in the transactions must accept what has come to be called a suboptimum situation. Suppose division A of a company makes spark-ignition engines, some of which it sells to division B, which assembles them into motor mowers for sale. The marginal cost to A of the engine is £2, and the optimum selling price £12. Apart from the engine, the marginal cost to B of the assembled lawn mower is £8, so that the total marginal cost of the lawn mower is either £10 or £20 according to whether division B pays £2 or £12 for the engine. If the price elasticity of demand for the company's brand of motor lawn mowers is 2, then the optimum price to B of the mower is either

$$10\frac{2}{2-1} = £20$$

$$\text{or} \quad 20\frac{2}{2-1} = £40$$

To the company, it is £20, which is unquestionably the rational selling price.

Suppose, now, that division B would sell 20 000 mowers a year when the selling price is £20. Then the value of K in Equation (D7.1) is equal to 20 000 x 20^2 = $8(10^6)$. At £40 each, therefore, and other things being equal, annual sales would be

$$Q = \frac{8\ 000\ 000}{40^2} = 5\ 000 \text{ mowers}$$

We can now make a comparison (Table T13.1) of the combined contribution (C22) that would be made by the two divisions to the company's annual profits and overheads at the two sets of prices.

Table T13.1

Selling price	Division	Unit contribution	No. of units a year	Annual Contribution	
£20	A	£2 − £2 = 0	−	£'000	£'000
	B	£20 − £10 = £10	20 000	200	200
£40	A	£12 − £2 = £10	5 000	50	
	B	£40 − £20 = £20	5 000	100	150

Difference, in favour of the marginal-cost transfer price
of £2 and engine   50

Provided that the rate of output of 20 000 mowers and engines lies within the normal capacity of the two plants (P17), the marginal-cost transfer price results in an increment of annual net profit of £50 000 to the company, compared with the fair-market-price transfer price.

Division A as a profit centre would have grounds for objecting to the marginal-cost transfer price; on the face of it, it would show an annual loss equal to its fixed costs (F8, also C26 and 28). However, as division·B would increase its own contribution from £100 000 to £200 000 as a result of a change from the fair-market-price transfer price to the marginal-cost transfer price, it should be willing to share the increment with division A. One-third of £200 000, based on the distribution of the total if the fair market price were charged, seems a reasonable share for division A. As this would exceed its original contribution of £50 000 a year by £16 667, the division should be well satisfied, and with an annual contribution of £133 333 to its credit, so too should division B. The necessary debits and credits could be arranged through a central contribution fund, which would form part of the interdivisional books of account, and to which all annual contributions made by the production divisions of the company would be credited.

It will be seen that the foregoing is highly relevant to make-or-buy policy (M1). It is common practice to decide make or buy by a comparison of annual costs only, as demonstrated in M1, but the foregoing makes it clear that the decision may affect the selling price, so that comparative annual gross revenues need to be taken into account, as well as costs.

Applications. Company integration; rationalizing interdivision trading.

T14   Treasury bill

A kind of official bill of exchange used by the Government and bill holders for a wide variety of purpose. It is the short-term counterpart of the long-term gilt-edged security. Treasury bills do not bear a rate of interest, but are issued at a discount and redeemed at par on maturity — usually 91 days after the date of issue. The rate of discount varies from time to time, in accordance with market conditions and the general level of interest rates. At the end of 1969, it stood at 7.65% a year, equivalent to an interest yield of 8.28% a year. It had fallen to 4.41% two years later and recorded a continuous rise in 1972, reaching 8.44% by the end of the year, when it gradually declined until the end of February 1973.

As a piece of paper, a treasury bill is about 9 inches long and 6 inches wide. It bears the dates of issue and maturity, and the amount payable on maturity. A bill for £10 000, for instance, reads:

London . . . 1973. This Treasury Bill entitles — — — — — — — — — or Order to Payment of ten thousand pounds at the Bank of England out of the Consolidated Fund of the United Kingdom on . . . th day of . . . 1973.

Denominations are £5000, £10 000, £25 000, £50 000 and £100 000.

There are two kinds of UK treasury bills: (1) market bills; (2) tap bills. Market bills are issued by weekly tender every Friday, the weekly issue ranging from £60 million to £300 million. Tap bills are 'on tap' to Government departments, which buy and sell them in order to maintain a fairly regular adequate cash balance and to minimize Government borrowing. As many tap bills find their way into the market, and so become market bills, the distinction resolves itself into bills held by Government departments and the Bank of England and those held outside, including those held by overseas central monetary authorities. Since market bills all now have a life span of 13 weeks, the theoretical amount outstanding at any time is equal to the issues during the previous 13 weeks. During the 13 weeks ending 30 June 1972, a total of £1 290 million was issued but, according to *Financial Statistics,* monthly, HMSO, the total of market bills held outside Government departments and the Bank on the same day amounted to £2 506 million. Whether the excess of actual stock over issued stock was due to a large net inflow into the market of tap bills is not clear. Some large holders may have retained their bills after the date of maturity. There are no days of grace.

To holders of treasury bills, they have two great virtues: they are a highly liquid asset and they yield a return. Many undertakings in the private sector (S12), as well as in the public sector, invest surplus funds in treasury bills — the banks, insurance companies, building societies and industrial companies. The current high market rates of discount make them very attractive as a secure store of value.

T15   Treasury method (TM)

One of the investment criteria, so called because it was invented by the UK Treasury and was in general use in Government departments up to about 1964. It converts the outlay on an asset to an annual capital charge (A11). The outlay on the asset is depreciated by the straight-line method, and interest is charged on half the outlay, on the assumption that an average of half the value of the capital is outstanding over the life of the asset. If the outlay on a wasting fixed asset is £20 000, the book life of the asset is five years, and the rate of interest is 7%, then the annual capital charge calculated by the Treasury method would be:

		£
Depreciation		
£20 000	=	4 000
Interest on half outlay		
£10 000 at 7%		700
Total annual capital charge		£4 700

There are theoretical grounds for supposing that the TM generally understates the annual charge and, therefore, overstates the net yield of the asset. The comparable full-discounting annual capital charge method is annual value (A12). For the example above, it gives a total annual capital charge of £4878:

Depreciation
Sinking fund (S18)   £
£20 000 at £0.1739 per £   =   3478
Interest on outlay
£20 000 at 7%   1400
Total annual capital charge   4878

As a rough-and-ready way of making a preliminary calculation, the Treasury method can scarcely be bettered.

Applications.   Investment appraisal.

T16   Trend ratios

A method of analysing the trend in a time series by calculating the ratio of the figure for period n to that for period $n-1$ throughout the series; sometimes called link relatives. A worked example of the method applied to sales forecasting is given in Table L14.1.

Applications.   Forecasting and time-series analysis generally.

T17   Trend coefficient

An arbitrary measure of the trend in a time series; originally designed to show the direction of trend where it is not discernible by inspection. It is equal to the ratio of the average of the series weighted by reference to an arithmetic progression, 0, 1, 2, 3, 4 ... n, to the unweighted average. Where the ratio is less than unity, the trend is downwards, where greater, it is upwards, and where it is unity, there is no trend. An example of each is given in Table T17.1. If the trend coefficient is to be used for trend-forecasting purposes, the series should first be converted to logarithms before the averages are derived, on the assumption that the trend is logarithmic and not linear.

The formula for forecasting the trend norm of the first forecast period is

$$F_1 = \frac{T_c \bar{Y}}{2 - T_c} \tag{17.1}$$

Table T17.1   Calculating the trend coefficient.

Period	weights	Example I Series	Example I Weighted series	Example II Series	Example II Weighted series	Example III Series	Example III Weighted series
1	0	3	0	6	0	4	0
2	1	1	1	2	2	1	1
3	2	3	6	3	6	4	8
4	3	2	6	1	3	3	9
5	4	6	24	3	12	3	12
Totals	10	15	37	15	23	15	30
Averages							
Unweighted		3		3		3	
Weighted			3.7		2.3		3.0
Trend coefficient		$\frac{3.7}{3} = 1.2333$		$\frac{2.3}{3} = 0.7667$		$\frac{3.0}{3} = 1.0000$	

where F_1 is the trend norm logarithm of forecast period 1, T_c the trend coefficient, and $\overline{Y}$ the unweighted average of the logarithms of the series. The trend norm logarithm of the middle period of the series is Y, so that we now have two trend norms, from which it is a simple matter to calculate the logarithmic period difference in the trend and so make it possible to calculate the trend norm of the logarithms for any given future period. The formula for calculating the period norm difference D is

$$D = \frac{F_1 - \overline{Y}}{N} \qquad (17.2)$$

where N is the number of periods from the middle period of the series to the first forecast period. Then

$$F_2 = F_1 + D, \ F_3 = F_1 + 2D, \text{ and } F_n = F_1 + (n-1)d \qquad (17.3)$$

As this method of trend forecasting has yet to be perfected, it is advisable to check the results by recalculating the trend coefficient of the actual series plus the forecast series. It should give the same answer as the actual series. If it does not, another method of trend forecasting, such as least squares (L7) or trend ratios (T16), should be considered.

Applications.   Forecasting.

T18   'Twopence off'

Probably the principal difficulty of measuring the effect on sales of this promotional method is its short-term nature. With a relatively new product having a name to make for itself, a less brittle method of sales promotion seems called for — below-the-line and routine advertising, for instance. 'Twopence off' may help, but what little evidence there is suggests that, alone, it lacks the force to raise the demand curve (P17) to a higher level and sustain it in that position. There is another point: if the recommended price is the optimum (R6), the cut price is a suboptimum, though, admittedly, the question 'optimum to whom?' must be asked.

u

U1 Uncertainty

If uncertainty were measurable, there would be no uncertainty. What is often measurable is the probability of an event (P28). When a coin is tossed, it is uncertain whether it will turn up heads or tails. There is a 0.50 probability that it will turn up heads, and a 0.50 probability that it will turn up tails. It is the probability that is measurable. If the coin is tossed ten times and turns up heads every time, the odds that it will turn up heads on the eleventh toss are still even, ie there is 0.50 probability. What has happened in the past does not affect the probability: it is the facts of the current situation that count, namely that the coin is unbiased, that it has two sides, one called heads and the other tails, and that it is tossed freely in the air to spin an unpredictable number of times.

These are objective measures of probability. So, too, are many of the statistical measures of probability, such as 0.68 for the standard deviation, and 0.95 for two standard deviations (N9, S24). However, there is a growing tendency to employ subjective estimates of probabilities in respect of the uncertainties involved in the alternative decisions open to a policy maker. The decision tree (D3) is an example.

U2 Unit cost

The ratio of the total annual cost of sales (C31) of a product to the number of items produced. Equation (C28.1) represents the total cost, namely

$$T = aQ + F$$

where T is the total annual cost, a the marginal cost, Q the quantity

made, and F the annual fixed cost. The unit cost is
$$T/Q = a + F/Q \qquad (2.1)$$

U3   Universe
The body of items from which a statistical sample is drawn, or which a sample is supposed to represent. More frequently referred to in the literature as a *population*. In statistical practice, a universe under research or analysis needs to be carefully circumscribed. The available universe may fall short of the ideal and, indeed, usually does. It is useful to prepare a list of weaknesses in the available universe to indicate the weaknesses that may exist in any sample drawn from it. If the available universe is incomplete, then since it is a principle of random sampling that every item in the universe should be given an equal chance of being selected, a sample drawn from it may be biased. Relevant entries are listed under S3.

U4   Unprovided renewals
In cost accounting, the total of the annual amounts of sinking-fund depreciation (S18) left unprovided when a wasting fixed asset is prematurely displaced (P22) to give way to a more technologically advanced asset designed to perform the same task. Whether, under a replacement scheme of this kind, the service of unprovided renewals is properly a charge against the scheme, or not, is a subject of some controversy. There are two alternatives open to a company: one is to write off the amount unprovided in its books of account, the other, to let it remain as a non-depreciable capital item. If the latter is chosen, the logical cost-accounting solution seems to be to deduct from the total amount that should be provided the amount so far notionally provided in the sinking fund. The amount so far provided is equal to the amount of annual provision invested annually at the sinking-fund rate of interest (A9). Suppose a machine tool with a book life of 10 years and a net replacement cost (R11) of £1000 is displaced at the end of the seventh year of its life, and suppose the sinking-fund rate of interest is 10%. Then:

	£
Amount to be provided	1 000

Annual sinking fund, 10 years at 10%
  £1000 at £0.06274 a year = £62.74

Amount provided in 7 years at 10%
Amount of £ a year   =   £9.487
£62.74 at £9.487   595
Uprovided renewals   £   405

This is charged to the replacement scheme as a non-depreciable part
of the capital outlay, ie annual interest is charged on it.

Applications.   Investment appraisal, efficiency auditing.

U5   Utility

Man cannot produce matter, but only utilities inherent in matter.
(Marshall, *Principles.*)

Marginal utility is the value to the purchaser, of an extra unit of any
given commodity or service; it is measured in terms of money. Thus,
a housewife who buys 100 lb of butter a year when the price is 20p
per lb, and 101 lb when the price is 19p, has a marginal utility of
butter of 19p for the 101st lb in a year. As the price falls, her
purchases increase and the amount of satisfaction or utility she
receives from the last unit diminishes. This phenomenon is referred
to as the law of diminishing utility.

The term *utility* is sometimes used by statisticians to refer to the
measure that a decision maker may give to each of a number of
potential outcomes of a decision, in order of preference, made by
referring to the expected monetary return and the risks involved. The
values range from 0 to 1.00, 0 being allotted to the least preferred,
and 1.00 to the most preferred outcome.

V1   Value added
Net output (N3).

V2   Value index
The ratio of the total sales value of output in a given year to that in a base year. See index numbers (I7).

V3   Value of money
The purchasing power of money (P39).

V4   Variable
1 *Adjective:* changeable.
2 *Noun:* factor, or a series of statistics relating to a factor, the latter being often referred to as a *random variable* (R3). Variables are of two broad classes: dependent (D9) and independent (I6). See V8.

V5   Variable costing
Analysing annual costs by reference to a variable factor or group of related factors, such as output (M4), plant capacity (P17), top management, manpower (E5, M2). Unqualified, the term may be defined as analysing costs by reference to the rate of output (C26). *Project variable costing:* estimating the net additional annual costs of providing and operating a new capital work.
    Applications.   Costing, cost analysis, cost forecasting, investment appraisal, efficiency auditing.

V6   Variability
Changeableness, generally in reference to annual costs by reference to some given factor (V5).

V7   Variance
The square of the standard deviation (S24), usually symbolized by σ^2:

$$\sigma^2 = \frac{\Sigma(x^2)}{n}$$

where Σ is the summation sign, x represents the mean deviations of X, the variable under review, and n is the number of items in the series of X. The variance provides a term in many statistical formulae. See A10.

V8   Variate
Same as the noun *variable* (V4). Attempts have been made, by using the term *variate* for the factor, and *variable* for the statistical series, to distinguish between the factor as such, and the statistical series that provides a measure of the factor.

V9   Variation, coefficient of
The ratio of the standard deviation of a series to the average of the series. It provides a measure of dispersion which makes general comparison possible. Both the standard deviation and the average are expressed in terms of the units of measurement of the basic series: tons, gallons, acres, £, yards, etc. The coefficient of variation has no such constraint: it is a purely statistical entity.

V10   Venture capital
Risk capital; money sunk in a project that may fail, less Government subsidies and insurance cover.

V11   Vicious Circle
Intercorrelation between two factors (C7 and 25) that tends to cause an undesirable growth in the strength of the factors.

V12   Visible Trade
That part of the country's trade with other countries that consists of the import and export of commodities, as classified in the Import

List (I3) and the Export List (E10). The balance of visible trade varies appreciably from year to year. In 1970 and 1971, there was an excess of exports (fob) over imports (fob) amounting to £12 million and £317 million, respectively. In 1972, there was an adverse balance of £692 million.

V13 Volume index

Quantity index (I7, Q4).

W1　Waste products

Substances including liquids and gases produced as incidentals in the process of production and distribution. Some, such as the swarf from an engineer's machine-shop, are saleable, when they may be described as by-products. Others, such as the fumes from chimneys and motor vehicles, are disposed of in the atmosphere, though there are legal limits in some localities to the rate at which fumes may be emitted from industrial establishments.

From the point of view of the firm, the most troublesome kind of waste product is that whose disposal is expensive. Sometimes it is profitable to convert waste into a saleable by-product by factory processing. As the cost of disposal as waste is saved, the net profit from such a project would be equal to the sales proceeds from the by-product plus the cost of disposal as waste less the total cost of processing. Care needs to be taken in costing. Disposal costs may include annual capital charges on plant or vehicles, which may not be entirely saved if the plant is closed down. Depreciation would be saved, but an interest charge on unprovided renewals (U4) may count as a debit to the new processing plant. At the programming stage, the annual capital charges on the new plant would count as an annual cost against the project, as would direct labour and materials costs (V5). But once the new plant has been provided and brought into use, only the product variable cost (M4) would count for costing purposes under a modern system of product costing.

W2　Weighted average

An average where each item is multiplied by a figure called a weight relating to the item itself in order that a representative average may be obtained. An extreme example is the average fare paid by passengers on a bus service. Tickets are issued in denominations of 5p, 10p and 15p. The unweighted average of these is:

$$\frac{5 + 10 + 15}{3} = 10p$$

but this is not by any means the average fare paid, which is equal to the ratio of the total receipts to the number of tickets issued, eg

	No.	Fare	Receipts
	10 000 at	5p	50 000
	4 000 at	10p	40 000
	2 000 at	15p	30 000
Total	16 000		120 000

$$\text{Average fare paid} = \frac{120\ 000}{16\ 000} = 7.5p$$

Here, the number of tickets issued is the weight, and the weighted average is obtained by dividing the sum of the products of item and weight by the sum of the weights:

$$\text{Weighted average of series } X = \frac{\Sigma(XW)}{\Sigma(W)}$$

where Σ is the sign of summation, and W is the weight.

Price and quantity index numbers (I7) are weighted averages. For simple worked examples, see P33, S38.

X1   *x*-axis

The horizontal axis of a diagram (Q1).

Y1   *y*-axis
The vertical axis of a diagram (Q1).

Y2   Yield
Profit from an investment in physical assets (eg machinery, plant, vehicles or buildings), or in financial assets (eg equities, fixed-interest bearing securities, a bank deposit or a building-society account). See Y3, also D21, E1.

Y3   Yield criteria
The terms in which the yield of an asset may be expressed. There are six criteria in all:

1   The annual yield in terms of money.
2   The annual yield in terms of money expressed as a perpetuity.
3   The rate of return on capital.
4   The rate of return on capital as a perpetuity.
5   The present value of the yield, NPV (N4).
6   The terminal valuation, TV (T3).

A simple numerical example will show that these six criteria are tautological: they merely provide different ways of expressing the same thing. Suppose an asset with a life of 10 years costs £1000 to provide and yields £100 a year over its life, the appropriate rate of interest (ie the company's cost of capital, C29) being 7%. Then we have the following six expressions of yield:

1   £100 a year net yield.
2   £49.2 a year in perpetuity.
3   10% of capital.
4   4.9% of capital in perpetuity.
5   £702.4 present value of net yield.
6   £1381.6 terminal valuation.

Conversion from one criterion to another is a simple process. The present value of £100 a year for 10 years at 7% is

£100 at £7.024 per £ = £702.4 (criterion 5)

This amount invested at 7% in an undated Government bond or similar security would yield

£702.4 at 7% a year = £49.2 a year (criterion 2)

Invested at 7% for 10 years, the sum of £702.4 would amount to the same as £100 a year at 7%:

£702.4 at £1.967 per £ = £100 at £13.816 per £
= £1381.6 (criterion 6).

And so on.

Oddly enough, although the six criteria are all interconvertible, they may give different rankings where alternative assets designed to perform the same task are under consideration. Two factors that may vary from asset to asset account for this phenomenon: the amount of the outlay, and the length of life. For instance, if one alternative asset costs £2000 instead of £1000, as in the example, then, although the money yield of £100 a year may remain the same, the rate of return on capital becomes 5% against 10%. If another asset has a life of 20 years instead of 10 years, then the present value of the net yield and the terminal valuation would be about three times greater; so, therefore, would the money yield in perpetuity. Two criteria, nos. 3 and 4, take care of differences in outlay, and four, nos. 2, 4, 5 and 6, take care of differences in asset life. Criterion 4 is the only one that takes care of both, so that where mutually exclusive projects are being ranked or otherwise compared, criterion 4, the rate of return expressed as a perpetuity, is the logical choice. (See A8, 9 and 11, C16, D16, I17.) Where the indivisible yield of a group of assets with differing lives is being sought, it seems best to use the annual-value method (A12) and to convert the money yield thus obtained into the rate of return in perpetuity.

Applications.　Investment programming and appraisal, efficiency auditing.

Y4　Yield variance

Profit variance (P37).

Z

Z1   Zero marginal cost

Marginal costs of zero value arise mainly in the service industries, particularly transport and electricity supply, where the unit of service is very small, eg ton mile, passenger mile, unit of electricity. A realistic approach for transport is to use the vehicle mile or the train mile as the unit of service, when the marginal cost would be greater than zero.

Z2   Zero time

A point of time defined as *present* in *present value* (P23); sometimes called *zero hour.*

   Applications.   Investment appraisal, efficiency auditing, pension-fund calculations.

Annotated Bibliography

General works in order of complexity

M. B. Brodie, *On thinking statistically,* Hutchinson, 1963. A brief, simple and lucid introductory text. No mathematics apart from arithmetic, and no index, but there are a few diagrams.

R. Loveday, *Statistics, a first course,* Cambridge University Press, second edition 1966. A somewhat less simple text devoted to the elementary concepts of statistical theory and practice. Mathematics: simple algebra.

Central Statistical Office, *Facts in focus,* Penguin, 1972. A compendium of interesting rather than useful basic statistics and index numbers.

S. Hays, *An outline of statistics,* Longman, eighth edition 1970. First published in 1937, this work has stood the test of time. Clearly written, with many numerical examples and diagrams largely devoted to economic and business problems. Mathematics: a very small amount of algebra, but a great deal of arithmetic.

W. R. Minrath, *Handbook of business mathematics,* D. Van Nostrand, 1959. Begins with simple arithmetic, on to algebra and the calculus. Then on to business applications: simple and compound interest, annuities, computers and operational research. Mathematics: fairly simple algebra, the chapter on the calculus is clearly written and not advanced.

W. M. Harper, *Statistics,* Macdonald and Evans, second edition 1971. A simple introduction to general statistics, mainly in note form for the use of students. Mathematics: elementary algebra.

International Tutor Machines Ltd, *Introduction to management statistics,* English Universities Press, 1967. A programmed text that needs patience to master. Many examples and exercises with answers. Mathematics: elementary algebra.

R. Brockington, *Statistics for accountants,* Gee, 1965. A useful elementary text designed for students studying for accountancy examinations. At least one definition is inaccurate: a chain-base index is not a formula for linking a new index series to an old one. Mathematics: elementary algebra.

G. L. Thirkettle, *Wheldon's business statistics and statistical method,* Macdonald and Evans, seventh edition, 1972. A good introduction to probability statistics and index numbers, but surprisingly weak on correlation and regression analysis. Mathematics: elementary.

F. E. Croxton, D. J. Cowden and S. Klein, *Applied general statistics,* Pitman, third edition 1968. A useful introduction to practical statistics for businessmen and economists. Mathematics: elementary.

E. J. Broster, *Management statistics,* Longman, 1972. Attempts to apply the methods of statistics, in tandem with microeconometrics, to the solution of a range of problems in the field of management accountancy. Mathematics: simple algebra with a modicum of the differential calculus.

M. J. Moroney, *Facts from figures,* Penguin, second edition, 1953. In view of the relatively low level of mathematics used, this work takes the reader a surprisingly long way towards an understanding of statistical theory and its practical applications to quality control and similar problems. Weak on index numbers and regression analysis. Mathematics: fairly simple algebra.

R. A. Fisher, *Design of experiments,* Oliver and Boyd, eighth edition, 1966. A work essentially designed for scientific research workers,

234

and the application of statistical methods in the pursuit of solutions of scientific problems. Mathematics: advanced.

B. W. Lindgren, *Statistical theory,* Macmillan, New York, second edition, 1968. Advanced academic theory and highly mathematical. Mathematics: advanced.

M. G. Kendall and W. R. Buckland, *A dictionary of statistical terms,* Oliver and Boyd, third edition 1971. Contains 2500 short definitions largely written in terms of mathematics and mathematicians' jargon. There are no numerical examples, and the mathematics are advanced. An invaluable work to mathematical statisticians.

M. G. Kendall, *The advanced theory of statistics,* Griffin. Volume 1, *Distribution Theory,* third edition, 1969; volume 2, *Inference and Relationships,* second edition, 1967. A standard work for mathematicians. It lies at the last frontier of theoretical statistics, and the mathematics are very advanced.

Specialized works
1 Business Ratios
C. A. Westwick, *How to use management ratios,* Gower Press, 1973. A work that more than lives up to its title: not only how to use but also how to choose and how to interpret. The author served for nine years with the Centre for Interfirm Comparison. His book promises to become the standard work on the subject, and is a *must* for top and middle managers in all kinds of companies: primary, service, and manufacturing; large and small. Over 400 ratios are examined and discussed. No mathematics.

L. G. Erskine, *Protect your profit margins,* American Management Association, 1965. Discusses 13 major ratios and their application to an actual but unnamed manufacturing company. Suggests removing seasonal variations by applying moving averages, and shows how to make 'basic norms' to provide a standard of comparison. No mathematics.

235

2   Costing
ICWA, *A report on marginal costing,* Institute of Cost and Works Accountants, 1961. An entirely non-mathematical treatment of the subject, and partly as a result, the work needs revision in some places to bring it up to date. Nevertheless, it remains an invaluable introduction to the subject.

ICWA, *Terminology of Cost Accountancy,* Institute of Cost and Works Accountants, 1966. Comprises about 240 brief definitions and one chart headed 'Variance analysis'. No mathematics.

R. Clugston, *Estimating manufacturing costs,* Gower Press, 1971. If one wished to provide a Latin tag for this work one could well use *multum in parvo.* It consists of no more than some 200 pages, yet it contains all, or nearly all, one might need on the subject. Well written and fully illustrated with diagrams and *pro formas.* Mathematics: a modicum of very simple algebra.

G. Shillinglaw, *Cost accounting: analysis and control,* Irwin, Homewood, Ill., 1967. An exhaustive work of some 900 pages, from the USA, with many numerical examples, charts and diagrams. Mathematics: a modicum of elementary algebra.

3   Index numbers
I. Fisher, *The making of index numbers,* Augustus M. Kelley, third edition, 1967. The standard work on the subject. Examines some hundreds of formulae, and finds them all wanting in one respect or another but concludes that the formulae of the Laspeyres-Paasche group are the best. The author was the first to propound Fisher's ideal index number, to which he gave his name.

4   Investment appraisal
A. M. Alfred and J. B. Evans, *Discounted cash flow,* Chapman and Hall, 1965. Favours and demonstrates the internal rate-of-return technique. Mathematics: simple arithmetic.

E. J. Broster, *Appraising capital works,* Longman, 1968. Intended as a complete guide to investment appraisal. Includes chapters on project variable costing, compounding and discounting, DCF techniques, pricing, forecasting and the main types of new capital works. Favours the annual-value technique of DCF. Mathematics: simple arithmetic.

ICWA, *The profitable use of capital in industry,* Gee, 1965. Favours the internal rate-of-return technique of DCF. An authoritative text by the Institute's technical staff.

A. J. Merrett and A. Sykes, *The finance and analysis of capital projects,* Longman, second edition, 1973.
A. J. Merrett and A. Sykes, *Capital budgeting and company finance,* Longman, 1966.
Of these two volumes, the latter is largely a simple, non-mathematical version of the former, which explains a number of concepts in terms of algebra. Makes a point of replacement works, compares the three main DCF techniques, and concludes that, on the whole, the internal rate of return is preferable to the other two.

NEDC, *Investment appraisal,* HMSO, 1965, revised 1967. The first edition favoured and demonstrated the internal rate-of-return technique, and the second, the net present-value technique, a change which critics of IRR described as a quick step in the right direction.

B. R. Williams and W. P. Scott, *Investment proposals and decisions.* Allen and Unwin, 1965. A report on the methods and procedures used at the programming stage in 14 large companies, including ICI, Beecham, Courtaulds, Shell-Mex, Uniliver, and Dunlop, based on one case from each. Discusses investment criteria in an appendix, and appears to favour the net present-value technique. A valuable contribution to the literature, but there is no index.

R. W. Wright, *Investment decision in industry,* Chapman and Hall, 1964. A somewhat philosophical approach to the problems of risk and uncertainty involved in the investment decision. No numerical examples. Discusses criteria briefly, without expressing any preference. Mathematics: fairly simple.

5 *Probability*
J. M. Keynes, *A treatise on probability,* Macmillan, 1921. A deeply

thought-out work based on the thesis that a probability statement expressing a relationship between two propositions may be as logical as a statement of logical consequence. The treatise is divided into five parts,, of which the last on 'The foundations of statistical inference' is of the greatest interest to statisticians. Keynes is concerned with that branch of statistics relating to inductive reasoning, which is bound up with probability. Mathematics: advanced.

P. G. Moore, *Risk in business decision,* Longman, 1972. An analytical and systematic approach, via probability theory, to the problem of risk. Many numerical examples, and exercises with answers. Mathematics: advanced.

E. Parzen, *Modern probability theory and its applications,* Wiley, 1960. A lucid exposition of the theory and practice of statistical probability, the treatment of random variables, and the law of great numbers. Mathematics: advanced.

6 Quality control
Dudding and Jennett, *BS600R quality control charts,* British Standards Institute.
Dudding and Jennett, *Control chart technique,* BSI.
These two works are concerned with the fundamental principles of making and using quality-control charts.

E. L. Grant, *Statistical quality control,* McGraw-Hill, third edition, 1964. A useful guide to the techniques of quality control. Mathematics: fairly simple algebra.

D. J. Desmond, *Quality control workbook,* Gower Press, 1971. A comprehensive and up-to-date text on the 'procedures for implementing statistical quality control', complete with *pro formas* and a valuable set of tables of control limits etc. Mathematics: simple algebra.

7 Regression analysis
E. J. Broster, 'Regression analysis for accountants' *Certified Accountant,* issues of December 1972 to March 1973. Demonstrates bivariate, multivariate, linear and curvilinear analysis, from model

building to calculating standard errors. Mathematics: elementary algebra.

R. A. Cooper, *Statistical models of economic relationships,* HMSO, 1971. A very brief introduction to regression analysis. Mathematics: simple algebra.

N. R. Draper and H. Smith, *Applied regression analysis,* Wiley, 1966. A comprehensive work devoted to the theory and application of least squares, and the use of matrix algebra in solving simultaneous equations. Mathematics: rather advanced.

M. Ezekiel and K. A. Fox, *Methods of correlation and regression analysis,* Wiley, 1959. A comprehensive and very readable text on several methods of regression analysis, ranging from the simple graphic method to least squares. Mordecai Ezekiel was in his day the leading authority on regression analysis, and his book *Methods of correlation analysis,* first published by Wiley in 1930, became the standard work on the subject. In both works there are numerous fully worked numerical examples. Mathematics: elementary algebra.

J. H. Heward and P. M. Steele, *Business control through multiple regression analysis,* Gower Press, 1972. Largely concerned with the linear analysis of labour time by the graphic and least-squares methods. Contains useful discussion on data preparation, the F test of significance, and the setting of control limits. A valuable contribution to the literature. Mathematics: elementary algebra.

8   Sources of statistics
Joan M. Harvey, *Sources of statistics,* Bingley, 1969. A handy volume to have for sources of environmental data.

9   Statistical tables
D. V. Lindley and J. C. P. Miller, *Cambridge elementary statistical tables,* Cambridge University Press, 1953.

K. Pearson, *Tables for statisticians and biometricians,* Cambridge University Press.

10   Systems analysis
H. D. Clifton, *Systems analysis for business data processing,* Business Books, 1969. A work designed to cover the whole range of processes for a computerizing project, from the analysis of existing systems to the creation, integration, implementation and appraisal of more efficient systems suitable for EDP. Minimum of jargon. Mathematics: none.

11   Trend forecasting
J. V. Gregg, C. H. Hossell and J. T. Richardson, *Mathematical trend curves: an aid to forecasting,* Oliver and Boyd, 1964. Monograph no. 1 of ICI's *Mathematical and statistical techniques for industry.* The authors belong to that school of thought which prefers the direct method of sales forecasting to the indirect, ie to extrapolate a time series of sales figures to arrive at forecast sales, rather than to predict sales by reference to a multiple-regression equation and forecasts of the several independent variables, which would include such important factors as changes in price and in advertising policy. The work is an outstanding exposition of the direct method. Mathematics: fairly advanced.

Index of Applications

Soc
HA
17
B76